Cephalopods from the Scientific Expeditions of Prince Albert I of Monaco

Cephalopods from the Scientific Expeditions of Prince Albert I of Monaco

VOLUME 2

Parts III and IV

Louis Joubin

Scientific Editors
Harald A. Rehder
Elizabeth K. Shea

General Editor
Clyde F.E. Roper

Foreword to the English Edition
Clyde F.E. Roper

Smithsonian Institution Libraries
and
The National Science Foundation
Washington, D.C.
1995

TT 81-52187
TT 81-52188

Céphalopodes provenant des Campagnes de la *Princesse-Alice* (1898–1910), 3e Serie
Imprimerie de Monaco, 1920
and
Contribution à l'étude des Céphalopodes de l'Atlantique Nord (4e Serie)
Imprimerie de Monaco, 1924

"The original French version of this document is available and can be purchased from the Service des publications du Musée océanographique de Monaco".

Translator: Dr. Indira Kohli
Editor: Dr. V.S. Kothekar

Library of Congress Cataloging-in-Publication Data

Joubin, L. (Louis), 1861–1935.
[Selections. English. 1989]
Cephalopods from the scientific expeditions of Prince Albert I of Monaco / Louis Joubin; scientific editors, Harald A. Rehder and Elizabeth K. Shea
p. cm.
Translated from French.
Includes bibliographies.
Contents: pt. 1. (1895) Contribution to the study of the cephalopods of the North Atlantic—pt. 2. (1900) Cephalopods collected during the expedition of the Princesse Alice (1891–1897) —pt. 3. (1920) Cephalopods collected during the expedition of the Princesse Alice (1898–1910)—pt. 4. (1924) Contribution to the study of cephalopods of the North Atlantic.
Supt. of Docs. no.: SI 1.2:C33/5/pt. 1–pt. 4
1. Cephalopoda—North Atlantic Ocean—Catalogs and collections. 2. Cephalopoda—Mediterranean Sea—Catalogs and collections. 3. Albert I, Prince of Monaco, 1848–1922 I. Rehder, Harald Alfred, 1907– . II. Shea, Elizabeth Keane, 1967– III. Smithsonian Institution Libraries. IV. National Science Foundation (U.S.). V. Title.
QL430.2.J862 1989 88-607952
594'.5'07404949—dc19 CIP

Printed at Baba Barkha Nath Printers, New Delhi, India

Contents

VOLUME 2

PART III (1920)

PART IV (1924)

PART III

Note by the Author

The printing of this memoir had been completed and the plates submitted to the printer in July 1914. The delivery of the finished plates could not take place until very recently because of the war, a fact that explains the long delay in the appearance of this volume.

L. Joubin

This fascicule was published and sent to the
Government of Monaco on June 30, 1920

Results of the Scientific Expeditions

Undertaken on His Yacht
by
ALBERT I
Crown Prince of Monaco

Published under his Supervision
with the help of
JULES RICHARD
Doctor of Science
In-charge of the On-board Zoological Work

FASCICULE LIV
Cephalopods Collected on the Expeditions
of the Princesse Alice (1898–1910)

Part 3

by *Louis Joubin*

with sixteen plates

Printed in Monaco
1920

Foreword

This is the third volume I have devoted to the study of cephalopods collected during the scientific cruises undertaken by HRH the Prince of Monaco. The first volume* was published in 1895 and the second in 1900.

I have the honor to request Prince Albert I of Monaco, Member of the Institute of France, to kindly accept my respectful appreciation of the confidence he has placed in me for 25 years.

I must also express all gratitude to my close friend Dr. Richard for the kind help he has given me in my work.

This volume contains descriptions of cephalopods captured during the concluding cruises of the *Princesse Alice,* before the initiation of the series of cruises made by the *Hirondelle II* in 1911.

The cephalopods described here were caught in almost the same seas as the earlier catches, i.e., the western Mediterranean, the Atlantic off the European coast, the coast of Morocco, the Canaries, the Azores, eastern part of the Sargasso Sea, the British Isles, coast of Norway, Spitzbergen, North Sea and the English Channel.

4 Some of the species mentioned in this memoir are the same as those described earlier. However, quite a number of additional forms have been described in much greater detail after having been described in various preliminary notes. Among these interesting new finds I first and foremost invite attention to several forms of *Cirroteuthis.* Only a very few species and specimens of these very curious octopods are known, frequently from only one; sometimes their earlier description was highly inadequate and imprecise. Several new species of this family were found for the first time in catches made by HRH the Prince of Monaco. I single out *Melanoteuthis lucens* among these exceptional cephalopods; this completely black, large cephalopod has several curious characteristics, one of which is sufficient to make it one of the most interesting cephalopods known: it has two large luminous organs on the dorsal side near the fins. This is the first time that such organs have been found in an octopod cephalopod, and it is also the first cephalopod to have these organs on the dorsal side. One could say that in invertebrates and marine fishes they are always ventral. This is a most interesting morphological feature.

The *Spirula* embryo is another highly peculiar animal or, to be more precise, a young *Spirula* emerging from the egg. It is the first time that a non-adult *Spirula* has been found. It is known that *Spirula* and *Nautilus* are the only living examples of cephalopods with a shell divided into chambers by septa traversed by a siphon. These cephalopods were abundant in the primary and secondary geological eras. For example, ammonoids constituted a preponderant part of the marine fauna of these epochs. All these forms abruptly disappeared at the end of the Cretaceous except the *Nautilus* and *Spirula.* These animals, particularly *Spirula*, are very rare and the anatomy of the *Spirula* is roughly known from descriptions of five to six specimens dissected more than half a century ago. Nothing is known about the development of these animals. The young form described here constitutes the first, though admittedly inadequate, document on the knowledge of the embryogeny of these animals. Nevertheless, the description has permitted us to understand some important points.

A description is also given of other embryos which are perhaps another species of *Spirula*, but their identification is doubtful.

*Included as parts I and II in volume one of this translation—General Editor.

I must also point out the interesting new species *Rossia caroli,* with some peculiarities particular to the morphology of this group.

Among the embryos and young described at the end of this volume, there are several that definitely belong to new genera and new species. Unfortunately, it is not advisable to base a new form on the description of its embryos. Cephalopods change greatly with age and a young form earlier believed to be a type of new
5 species may belong to a well known adult animal. These young cephalopods do not always have sufficiently distinctive diagnostic characters. As such, I have been compelled more than once to attribute only a generic name to them. A series of various stages of embryos is required for more precise determination and these are not always available. The photographs that I give of the embryos will be useful in the future when further cruises will yield more stages which can be intercalated in the series. This sort of handicap is further aggravated by the fact that the young do not generally preserve well and become deformed if they are not carefully placed in special reagents immediately after their capture. The job of sorting them from plankton is extremely time consuming and usually cannot be completed while the plankton is fresh. In these circumstances one is thus forced to make imperfect documents.

I must note the importance of the captures made thanks to the large plankton nets devised by Dr. Richard and lieutenant-commander Bourée. These enabled a large quantity of water to pass through at high or low rates of speed, and many cephalopods were caught which would have otherwise escaped by swimming away. The catches made by these nets on the yacht *Hirondelle II* will be given in volume four of the series on cephalopods, where the importance of these methods from the zoological point of view will be made apparent.

One of the cruises of the *Princesse Alice* was to Spitzbergen. The waters of these cold seas are particularly poor in cephalopods; it is, therefore, not surprising that only a few reports on them have so far been published. It is also the main cause of delay in publishing this volume. There is, however, another reason which I must mention. This volume was almost completed in the middle of 1912 when HRH the Prince of Monaco conferred on me the great yet onerous honor of organizing the IX International Zoological Congress at Monaco. The work in connection with this enormous task, the publication of the memoirs and the deliberations of the Congress took up so much time that my professional duties languished until the end of 1913 at which time I once again took up the task of editing my work. The dreadful events that took place during printing further retarded publication.

I would like to make a few comments here on the subject of names that I believed could be attributed to some cephalopods.

Since the establishment of rules for the creation of generic and specific names of animals, one would have assumed that a method had been found to remove uncertainties, confusion and all sorts of inaccuracies which confront systematic zoology. None of these has so far been removed because the framers, with the best of intentions, have pronounced as a fundamental rule, the requirement that priority should be the first thing to establish for the adoption of a name. As a consequence, it followed that names such as *Actinia,*
6 *Holothuria* and a thousand others, accepted throughout the world for a century, entered into science and used in all treatises on zoology, were marked as not having priority by those who combed the libraries. There are people dedicated to rummaging through old papers, justifiably forgotten by all, who have transformed zoology into a veritable Tower of Babel under the pretext of respecting the rule of priority; many zoologists did not wish to accept these useless changes; they preferred to maintain the rule of usage rather than adopt the rule of priority with all the radicalism of its absurd consequences. They felt that there would be no guarantee against others ferreting through libraries of even older papers, which would oblige them to again change the generic and specific names and thus increase the confusion.

I side with the zoologists protesting against the abuses of the rule of priority. I have maintained the name *Octopus* for the devilfish. For a century now, all zoologists have used this name in their memoirs and treatises. I am convinced that my eminent colleague and friend, Dr. Hoyle, has harmed zoology in general and the study of Cephalopoda in particular by exhuming an old 1784 text in which the poor devilfish is called *Polypus.* This name causes confusion; one would not know whether it would mean a hydroid, a coral or something else. What benefit has this discovery brought to zoology? None. And Hoyle may some day find an even earlier name for the devilfish and we will be obliged to change the name again. For these various reasons I maintain, as before, the genus *Octopus.* The entire world will understand what I wish to say. I also maintain the genera *Eledone* and *Eledonella.* I think it is completely pointless to give these animals a baroque new name as has been proposed under the pretext of priority.

I have explained my point of view at length so that I cannot be accused of negligence in preserving the names accepted by all zoologists for more than a century.

In closing this introduction, I must express my profound gratitude to Dr. G. Pfeffer, the learned naturalist of the Museum of Hamburg for his kindness. At different times I have sent him specimens of young cephalopod about which I had doubts. We exchanged lengthy letters on this subject and, since Dr. Pfeffer is the highest authority on such delicate organisms, I have almost always accepted his opinion.

Classification

Cephalopoda

SUBORDER OCTOPODA LEACH, 1818

Division A. LIOGLOSSA Lütken, 1882

Family I. PTEROTI Reinhardt and Prosch, 1846

Genus *Cirroteuthis* Eschricht, 1836

1. *Cirroteuthis umbellata* P. Fischer, 1883
2. *Cirroteuthis megaptera* Verrill, 1885
3. *Cirroteuthis grimaldii* Joubin, 1912

Genus *Melanoteuthis* Joubin, 1912

4. *Melanoteuthis lucens* Joubin, 1912

Division B. TRACHYGLOSSA Lütken, 1882

Family II. PHILONEXIDAE d'Orbigny, 1838

Genus *Tremoctopus* Delle Chiaje, 1830

5. *Tremoctopus violaceus* Delle Chiaje, 1830

Family III. ALLOPOSIDAE Verrill, 1881

Genus *Alloposus* Verrill, 1880

6. *Alloposus mollis* Verrill, 1880

Family IV. OCTOPODIDAE d'Orbigny, 1838

Genus *Octopus* Lamarck, 1799

7. *Octopus vulgaris* Lamarck, 1799
8. *Octopus arcticus* Prosch, 1849
9. *Octopus* sp.

8 **Genus *Eledone* Leach, 1817**

10. *Eledone cirrhosa* Leach, 1817

Genus *Eledonella* Verrill, 1884

11. *Eledonella diaphana* Hoyle, 1885
12. *Eledonella* sp.

SUBORDER DECAPODA LEACH, 1818

Division A. MYOPSIDA d'Orbigny, 1845

Family I. SEPIOLINI Steenstrup, 1861

Genus *Heteroteuthis* Gray, 1849

13. *Heteroteuthis dispar* (Rüppell) Gray, 1849

Genus *Rossia* Owen, 1834

Subgenus *Franklinia* Norman, 1890

14. *Rossia caroli* Joubin, 1902
15. *Rossia glaucopis* Lovén, 1845

Subgenus *Rossia*

16. *Rossia macrosoma* (Delle Chiaje) d'Orbigny, 1839
17. *Rossia palpebrosa* Owen, 1835

Family II. SEPIARII Steenstrup, 1861

Genus *Spirula* Lamarck, 1801

18. *Spirula* sp.
19. *Spirula* sp.

9 **Genus *Sepia* Linné, 1766**

20. *Sepia elegans* d'Orbigny, 1826
21. *Sepia orbignyana* Férussac, 1826

Division B. OEGOPSIDA d'Orbigny, 1845

Family I. ONYCHOTEUTHIDAE Gray, 1849

Genus *Onychoteuthis* Lichtenstein, 1818

22. *Onychoteuthis banksi* (Leach) Férussac, 1817

Genus *Chaunoteuthis* Appellöf, 1890

23. *Chaunoteuthis mollis* Appellöf, 1890

Family II. ENOPLOTEUTHIDAE G. Pfeffer, 1900

Genus *Abraliopsis* Joubin, 1896

24. *Abraliopsis pfefferi* Joubin, 1896
25. *Abraliopsis* sp.

Genus *Thelidioteuthis* G. Pfeffer, 1900

26. *Thelidioteuthis alessandrini* (Vérany), 1851

Genus *Pterygioteuthis* P. Fischer, 1896

27. *Pterygioteuthis giardi* P. Fischer, 1896

Genus *Enoploion* G. Pfeffer, 1912

28. *Enoploion eustictum* G. Pfeffer, 1912

10 Genus *Octopodoteuthis* Rüppell, 1844

29. *Octopodoteuthis sicula* Rüppell, 1844

Family III. HISTIOTEUTHIDAE G. Pfeffer, 1912

Genus *Stigmatoteuthis* G. Pfeffer, 1900

30. *Stigmatoteuthis chuni* G. Pfeffer, 1912

Genus *Calliteuthis* Verrill, 1880

31. *Calliteuthis* sp.

Family IV. BENTHOTEUTHIDAE G. Pfeffer, 1912

Genus *Benthoteuthis* Verrill, 1885

32. *Benthoteuthis megalops* Verrill, 1885
33. *Benthoteuthis* sp.

Genus *Ctenopteryx* Appellöf, 1890

34. *Ctenopteryx* sp.

Family V. BRACHIOTEUTHIDAE G. Pfeffer, 1912

Genus *Tracheloteuthis* Steenstrup 1881

35. *Tracheloteuthis* sp.

Family VI. OMMATOSTREPHIDAE G. Pfeffer, 1900

Genus *Ommatostrephes* d'Orbigny, 1835

36. *Ommatostrephes sagittatus* Lamarck, 1799

11 Genus *Illex* Steenstrup, 1880

37. *Illex illecebrosus* (Lesueur) d'Orbigny, 1821

Genus *Todaropsis* Girard, 1890

38. *Todaropsis eblanae* (Ball), 1841

Genus *Stenoteuthis* Verrill, 1880

39. *Stenoteuthis bartrami* (Lesueur), 1821

Genus *Rhynchoteuthis* Chun, 1903

40. *Rhynchoteuthis* sp.

Family VII. CHIROTEUTHIDAE Gray, 1849

A. Subfamily Chiroteuthinae G. Pfeffer, 1912

Genus *Chiroteuthis* d'Orbigny, 1839

41. *Chiroteuthis* sp.

Genus *Doratopsis* de Rochebrune, 1884

42. *Doratopsis* sp.

B. Subfamily Mastigoteuthinae G. Pfeffer, 1912

Genus *Mastigoteuthis* Verrill, 1881

43. *Mastigoteuthis magna* Joubin, 1913
44. *Mastigoteuthis* sp.
45. *Mastigoteuthis* sp.

12 Family VIII. CRANCHIIDAE G. Pfeffer, 1912

A. Subfamily Cranchiinae G. Pfeffer, 1912

Genus *Leachia* Lesueur, 1821

46. *Leachia cyclura* Lesueur, 1821

Genus *Cranchia* Leach, 1817

47. *Cranchia scabra* Leach, 1817

B. Subfamily Taoniinae G. Pfeffer, 1912

Genus *Taonius* Steenstrup, 1861

48. *Taonius pavo* (Lesueur), 1821

Genus *Galiteuthis* Joubin, 1898

49. *Galiteuthis armata* Joubin, 1898

Genus *Teuthowenia* Chun, 1910

50. *Teuthowenia megalops* (Prosch), 1849

Genus *Hensenioteuthis* G. Pfeffer, 1900

51. *Hensenioteuthis joubini* G. Pfeffer, 1900

Genus *Bathothauma* Chun, 1906

52. *Bathothauma lyromma* Chun, 1906

Genus *Fusocranchia* Joubin, 1913

53. *Fusocranchia alpha* Joubin, 1913

13 # Descriptive Part

SUBORDER **OCTOPODA** LEACH, 1818

Division I. LIOGLOSSA Lütken, 1882

Family I. PTEROTI Reinhardt and Prosch, 1846

Genus *Cirroteuthis* Eschricht, 1836

Cirroteuthis umbellata P. Fischer, 1883

1883. *Cirroteuthis umbellata* P. Fischer (**13**), p. 402
1900. *Cirroteuthis umbellata*, L. Joubin (**37**), p. 21, pl. I.
?1904. *Stauroteuthis hippocrepium* Hoyle (**27**), p. 6, pl. II and III

Two specimens of this attractive cephalopod were caught in the middle of the Gulf of Gascony.

Expedition of 1910: Specimen A. Stn. 2986, depth 4870 m.

Total length of this specimen approximately 20 cm including the arms. Body gray. Crown of arms and membranes (web) between them deep reddish-brown.

A nodule is present at the junction of the web and each arm. This web is less developed between the ventral arms than between the dorsal and intermediate arms; it diminishes gradually from the dorsal median line to the ventral.

Some measurements of the specimen are given below:

		mm
Length from tip of arm to the end of mantle	..	205
Length of 1st arm (dorsal)	..	137
Length of 2nd arm	..	135
Length of 3rd arm	..	115
Length of 4th arm (ventral)	..	112
Length of web between 2 dorsal arms	..	92
Length of web between arms 1 and 2	..	81
Length of web between arms 2 and 3	..	59
Length of web between arms 3 and 4	..	48
Length of web between the two ventral arms	..	36
14 Distance from tip of one fin to the other	..	120
Length of mantle from below the funnel	..	42

The lengths of the arms and web were measured from the mouth. There are 54 suckers per arm; the first four are very small; the following 26 are noticeably equal in size; the remaining become progressively smaller to the tip of the arm.

The cirri are so small that they can hardly be discerned in the last 3 or 4 cm of the arm.

The cartilage is horseshoe-shaped with the two horns directed toward the head.

The slender fins are colorless except for a slight red-brown tinge on the posterior margin.

Mantle opening rounded, markedly larger than the funnel which extends beyond it for a length of 17 mm.

Specimen B. Stn. 2994, depth 5,000 m.

This second specimen is in rather bad condition: the web disintegrating and almost impossible to measure exactly. Colors similar to specimen A.

		mm
Length from tip of arm to the end of mantle	..	235
Length of 1st arm (dorsal)	..	165
Length of 2nd arm	..	148
Length of 3rd arm	..	132
Length of 4th arm (ventral)	..	121
Distance from tip of one fin to the other	..	126
Length of mantle from below the funnel	..	47

The projecting part of the funnel is approximately 19 mm. Dorsal arm has 60 suckers, the first five from the mouth are very small. Similarly, the last ten at the tip of the arms are very small; it is uncertain whether some are missing. Ventral arm has 56 suckers.

I think these two cephalopods must be assigned to *Cirroteuthis umbellata* Fischer (**13**).

In an earlier memoir (**37**) I mentioned that I did not find the extended shape of the mantle cartilage adequate for a division of the old genus *Cirroteuthis* into two genera as was done by Hoyle. According to this author, the genus *Cirroteuthis* includes species in which the cartilage is saddle-shaped, whereas the genus *Stauroteuthis* includes species in which the cartilage is horseshoe-shaped. This feature is extremely difficult to determine accurately since generally only one or two specimens become available for dissection. Besides, this point can be deferred until more individuals have been examined. Finally, the cephalopods of these two
15 genera resemble each other very closely; all their other features, except the one mentioned above differ so little that there are really no adequate grounds to justify a division of the old genus *Cirroteuthis* into two genera, *Stauroteuthis* and *Cirroteuthis* sensu stricto.

For these reasons I leave these cephalopods under the name of *Cirroteuthis umbellata,* despite their horseshoe-shaped cartilage.

Moreover, I suggest that the sole specimen of *Stauroteuthis hippocrepium* Hoyle, studied by this author, must also be assigned to *Cirroteuthis umbellata* Fischer.

The cephalopod studied by Hoyle was caught at 1,823 fathoms near Malpelo Island in the open sea of the Gulf of Panama in the Pacific. Unfortunately it was in very bad condition and the author could not study it adequately. The reconstructed illustration does not seem to differ much from the specimens I have studied, or differ from the original type of Fischer's species. I have doubts about synonymizing the two species since I have no information other than Hoyle's publications and I have not seen his cephalopod. However, it is not unlikely that the two species are synonymous, with the priority assigned to Fischer's species.

Cirroteuthis megaptera Verrill, 1885
(Pl. I, Figs. 1 to 4)

1885. *Cirroteuthis megaptera* Verrill (**89**), p. 405
1903. *Cirroteuthis richardi* Joubin (**41**)

Expedition of 1901: Stn. 1150, depth 3,890 m.

A single specimen of this *Cirroteuthis* was caught and a water-color sketch was made while it was fresh. In a preliminary note on all the catches from this cruise, I observed that this specimen differed sufficiently from *C. megaptera* of Verrill to be placed in a separate species that I named *C. richardi.* A more detailed study of this cephalopod has led me, despite a few differences, to assign it to the species of Verrill.

The individual caught by the Prince of Monaco is bright carmine between the arms and on the buccal surface of the umbrella, and paler on the body and fins. The suckers are slightly lighter than the membrane surrounding them.

Figure 1, Plate I shows a photograph of the specimen after it was preserved in formalin. The length of the animal in that state, with the arms coiled, is 60 mm. If the arms were uncoiled it would probably be necessary to add 20 to 25 mm to the length, but I did not wish to undertake this operation since the skin was delicate and already in shreds at the base of the arms and on the body.

16 The span of the fins in their normal position, as seen in the photograph, is 48 mm between the two tips.

The two figures given by Verrill (**89**, Pl. XLIII) are too diagrammatic. I have reproduced the photographs in Plate I to give a more exact idea of the proportions of the various organs, suckers, cirri, fins, etc. It can be seen that the funnel is longer than that shown in Verrill's illustration; the same applies to the cirri. Since the skin is ruptured at the base of the arms, the cylindrical and transparent axes which form the basic structure can be seen.

Verrill mentions muscular organs found on the external surface of the arms, if not on all then at least on some of them; these organs probably support or reinforce the fin. I did not find either these or anything resembling them, in my specimen. Due to the absence of these organs I initially decided to erect a new species for this animal. However, I have thought that their absence might be due to the poor condition of Verrill's cephalopod and also mine, and that it would be more prudent to await further proof before separating *C. megaptera* and *C. richardi.*

As already mentioned, the genus *Cirroteuthis* has been divided into two genera: *Cirroteuthis* sensu stricto and *Stauroteuthis.* Both have interbrachial membranes in the form of an umbrella. However, the supporting cartilage of the fins is saddle-shaped in *Cirroteuthis* and horseshoe-shaped in *Stauroteuthis.* Based on this, Hoyle (**25**) placed *C. megaptera* of Verrill in genus *Cirroteuthis* sensu stricto. I have observed that without a doubt the cartilage is perfectly horseshoe-shaped. It is highly developed, probably in consonance with the large fins. *Cirroteuthis megaptera* should accordingly be placed in *Stauroteuthis*; I have, however, already explained why I do not accept this feature as adequate to separate the two genera. Pending further proof, I consider this specimen belonging to the genus *Cirroteuthis,* despite its horseshoe-shaped cartilage.

Cirroteuthis grimaldii Joubin
(Pl. II, III, IV and XV, Fig. 1)

?1903. *Cirroteuthis grimaldii* L. Joubin (**41**), p. 100–102
1912. *Cirroteuthis grimaldii,* L. Joubin (**44**), p. 1 to 13

Expedition of 1902: Stn. 1334, depth 1,900 m. To the NNW of Fayal, Azores. One specimen in good condition.

Expedition of 1903: Stn. 1450, depth 1,804 m. Off the Gulf of Gascony. This second specimen is in very bad condition.

The following description pertains solely to the larger of the two specimens which is in an excellent
17 condition. By comparison, the second was in such a sorry state that I had doubts as to what species it should be attributed. I tried to make several anatomical observations which further mutilated the animal. The second part of this section will therefore be devoted to the description of the structure of the second specimen.

In a preliminary note published in the Proceedings of the Academy of Sciences (**41**) I reported a new species of *Cirroteuthis* which I named *C. grimaldii.*

I photographed it shortly after its capture, and the illustrations in this work have been based mainly on these pictures.

The overall shape of the body is that of a large egg. This is because the ends of the arms are folded over the umbrella enclosing the opening and continuing the ovoid shape of the mantle region of the animal (Pls. II and III).

The complete absence of constrictions, which in all other forms separate the arms from the head and the head from the body, accentuates the peculiar shape of this cephalopod. Here all looks the same and no distinct regions can be demarcated in this ovoid form. I must state that the handling necessary to study this soft and water-saturated specimen gradually but noticeably changed its appearance. As a result, we could distinctly distinguish the musculature of the arms not otherwise visible in the fresh animal. The strength of this musculature can be assessed as it occupies by itself at least three-fourths of the total body mass.

It is interesting to observe that this animal whose skin is soft, flabby and puffy is, to the contrary, hard and solidly muscular when palpated along its cutaneous cover which is medusoidal in consistency.

Two fins, small in relation to its size, are on the posterior part of the body. Their size is noticeably less proportionally than that of the fins of other cephalopods of the same genus (Pls. II and III).

The only part of the eyes that is visible is the wrinkled opening of the circular eyelid which reduces them to a small hole. But, on palpation, two large globes can be felt buried under the skin. They are very far removed from each other, because I found a distance of 111 mm between the two eyelid openings.

On the ventral side, a short, conical funnel projects from a mantle opening just large enough to permit its emergence. The extreme reduction of this opening is highly characteristic of this *Cirroteuthis.* This peculiarity contributes the most toward giving the entire body the appearance of a uniform sac without any obvious separation of regions.

The maximum length of the animal is 217 mm, measured from the tip of the longest arms to the middle of the space between the fins. Its maximum width is 132 mm at the upper third of the umbrella; between
18 this level and that of the eyes the width is 127 mm; it is 111 mm between the two eyelid openings, and 96 mm between the upper insertions of the two fins.

The distance between the tip of the funnel and the free edge of the umbrella is 90 mm. I measured 191 mm along the dorsomedial line from the base of the body to the free edge of the umbrella.

Finally, the free part of the funnel is 27 mm long and emerges from the mantle cavity through an elliptical opening 21 mm in diameter.

The umbrella is formed by the arms and the membranes between them. The free part of the arms is shorter than the webbed part. It is approximately two-fifths free with three-fifths webbed, but this ratio varies slightly for each arm.

The umbrella is shaped like a deep funnel with such thick walls, due to the puffy condition of the tissues, that the musculature of the arms disappears even though it is very strong. It is not visible either inside or outside this wall where the arms have a proper shape beginning only at the free edge of the umbrella. This is not at the same level everywhere in relation to the base surrounding the mouth. The distance between the mouth and the intervals of the arms is 99 mm for the ventral side and 132 mm for the dorsal side, with 113, 125 and 132 mm for the indentations of the intermediate arms.

All these measurements were difficult to take due to the softness of the surface tissues and the tendency of the specimen to flatten and become deformed when handled out of water.

The overall color is rather difficult to describe. When the specimen was freshly preserved it was mottled with numerous reddish-brown spots; these gradually faded and the whitish background of the skin became predominant. The brown spots are reduced to sparsely scattered dots. A slightly more violet color is uniformly persistent inside the umbrella where it forms a continuous coating on the skin. The surface of the arms inside the umbrella is slightly lighter than that of the membranes between them.

The funnel is slightly darker than the skin of the body: it is blackish-brown, more red at its base, more brown toward its summit where the opening is ornamented by a whitish ring.

The arms do not differ greatly in length. They are not all complete, because 2 to 10 mm of the tip, being very slender and fragile, is broken off in each arm. The suckers of these filiform extremities are microscopic.

19 The lengths given below were measured from the mouth by placing the end of a graduated scale in the buccal opening.

1st arm (dorsal)	right	214,	left	205 mm
2nd arm	,,	212,	,,	217 mm
3nd arm	,,	202,	,,	197 mm
4th arm (ventral)	,,	191,	,,	191 mm

Approximately 1 cm had been broken off from the tip of the 1st left dorsal arm.

There is nothing to note on the outer, webbed surface of the arms. From the point where they project from the web they are very slender and are traversed along their edges by prolongations of the interbrachial membrane, which thins down almost up to the filiform tip.

The inner surface of the webbed part of the arms is flat, is traversed along its middle by the line of suckers and is bordered by a double line of cirri. This surface is smooth and contrasts with the longitudinally folded surface of the interbrachial membrane. In the middle of the arm the space between the two rows of cirri is 13.5 mm wide, each sucker is 3.5 mm in diameter and there are two 5 mm wide flat spaces between the cirri and the sucker. Thus, the measurements from one edge of the arm to the other at the level of the twentieth sucker are: cirrus 1 mm, flat space 5 mm, sucker 3.5 mm, flat space 5 mm, cirrus 1 mm. This proportion is maintained without variation over two-thirds of the length of the arms. It only differs near the mouth where the suckers are larger, and close to the tip where they are very much smaller. Even around the mouth there are two or three smaller suckers deeply buried in the tissues.

The cirri, characteristic of *Cirroteuthis,* are normal in their placement, but are very small and very far removed from the suckers. In this they differ from the situation seen in other species of the same genus.

They commence from the third or fourth sucker counting from the mouth, but are so small that it is extremely difficult to distinguish them. They progressively increase in size to a maximum of 3 mm in length toward the twentieth sucker. Then they decrease and are no longer discernible on the terminal filiform portion of the arms. These are the same color as the web but slightly lighter.

The suckers, in a single row on each arm, can be described in the following manner: From the mouth onward there are three small suckers deep in the epidermis, suddenly they become large, 6 to 7 mm, until the seventh. From the eighth to the thirty-sixth they are almost identical in size, about 3.5 mm. They progressively become smaller from the thirty-seventh to the forty-eighth. From then to the tip of the arm they are microscopic and can be counted only under a high-power magnifying lens. Here and there some may be missing, particularly
20 at the tip. I counted 86 suckers on the fourth left ventral arm; this was the maximum number. I found 70 to 81 suckers on the other arms, but some were definitely missing.

The suckers are deeply embedded in the skin and only the peripheral rim of their opening rises above the epidermis. The circular opening gapes and the groove of the rim is striated with fine radial lines. I could not distinguish any peculiarities on the interior of the suckers. I do not know whether the four large suckers which are approximately 6 mm in diameter, which is nearly double the size of the subsequent suckers, are due to a sexual difference.

The fins are usually short, but the skin to which they are attached is so loose that it is difficult to take exact measurements; even to place a compass on this skin deforms it and the measurements vary. With these reservations, I consider the fins to be shaped like an oval disk inserted at the posterior edge of the body where the contour passes from the dorsum to the side. The insertion is oblique so that the distance between the tip and the anterior edge of the insertion is 30 mm, whereas it is 45 mm between this same point and the posterior edge of the insertion. The maximum length of the swimming disk is 17 mm; its anterior edge is soft, thin and flexible; its posterior edge is stiffer and the shape is a continuous curve without inflections.

The fins have internal muscles visible within the more delicate part of the disk. They can be felt within the thickness of the body beyond the insertion of the fins. The muscles are attached to a semicircular cartilage which forms the internal skeleton. Since I did not dissect the specimen I could give only a very approximate drawing (Pl. XV, Fig. 1).

Both the fin muscles and their supporting semicircular cartilage form the solid part of the soft visceral mass. The visceral mass lies in the 96 mm space lying between the two upper insertions of the fins, and is deeply embedded in puffy integumental tissue supported by the skeletal, semicircular cartilage whose diameter is approximately 4 cm.

As I have already mentioned, since I did not wish to destroy this beautiful specimen I did not dissect it. The description of this cartilaginous ring is thus the result of an examination by palpation. It has exactly the shape of a semi-circle. Its edge is marked by a distinct ridge noticeable throughout its length. It extends at both ends into a slender cartilaginous needlelike projection approximately 2 cm in length and rising slightly towards the funnel. The fin muscles are inserted at the two ends of the semicircle.

The shape of the cartilage has a certain importance since it was thought possible to split the old *Cirroteuthis* into two new genera on this basis: the true *Cirroteuthis* with a saddle-shaped cartilage and the *Stauroteuthis*
21 with a horseshoe-shaped cartilage. I must confess that I do not attach such importance to the shape of this cartilage as to make it a criterion for distinguishing the two genera. In both cases it is semicircular, flat and dilated in one, and a half-ring in the other. These are at most specific differences and will probably lose all exclusiveness when intermediate forms are studied.

However, based on its cartilage my *Cirroteuthis grimaldii* should be named *Stauroteuthis,* but I shall leave it for the present in the genus *Cirroteuthis* until I have found a more precise differentiation between the two genera.

Based on its external shape, *Cirroteuthis plena* Verrill, 1885, seems to be most closely affiliated to *C. grimaldii,* but differs from it in several important features.

The small eyes of the American species are placed much closer to the arms; the fins are larger and inserted very high, not far from the base of the arms. The mantle opening is a transverse slit. In *C. grimaldii,* the fins are situated right at the back and they are on the posterior limiting line of the mantle. The eyes are very large. It has up to 86 suckers; *C. plena* has only 55 which are slightly different in size. The mantle opening is round.

Based on the shape of the mantle cartilage, as mentioned earlier, Hoyle placed the *C. plena* of Verrill in the true genus *Cirroteuthis,* whereas the *C. grimaldii* is included in his genus *Stauroteuthis* because of the shape of its cartilage.

Genus *Melanoteuthis* Joubin, 1912

Melanoteuthis lucens Joubin, 1912
(Pl. V, VI, VII, Figs. 1–3 and Pl. XV, Figs. 2–7)

1912. *Melanoteuthis lucens* Joubin (**43**), No. 220

Expedition of 1905: Stn. 2118, depth 3,465 m. Sargasso Sea. Stn. 2168, depth from 0 to 2,000 m. To the south of the Azores.

The following description is based on the two specimens. One is larger and its external features have been described, whereas the other is very much smaller and some anatomical studies have also been conducted on it. A colored sketch of the larger specimen was made by M. Tinayre and is given in Plate V. The smaller specimen was in a very bad condition and I could not make any reproduction of it, not even a photograph.

22 It might be remarked that these animals were almost absolutely black when they were fresh but their immersion in formalin diluted with sea water almost completely destroyed the pigmentation and only some traces have remained. Therefore, they do not have the same appearance as at the time of their capture. M. Tinayre's water-color sketch is, however, a valid record from which the features can be described. Moreover, the particularly soft, gelatinous, and hydrated body has shriveled and its general shape has greatly changed. It is wrinkled, folded and semi-transparent, particularly between the funnel and the base of the arms.

The only pigmentation that has remained is the dark violet-brown within the umbrella, particularly around the mouth. The other traces of pigmentation form lines or patches around the luminous organs and mantle opening.

Like the larger specimen, the smaller one also had only traces of black pigmentation; it was damaged to a considerable extent and one of the luminous organs was missing. Although its extended length was only 15 mm, I was able to dissect it, noting two interesting points.

The examination of this cephalopod was extremely difficult because of the soft texture of the tissues. Merely handling it and returning it to the jar of formalin, removed some of the liquid from the tissues. Thus, it reverted to its original shape. Moreover, the epidermis had become detached almost everywhere while in the net and during preservation in the formalin. Photographs of the animal in this condition could not have provided any useful information.

It is quite understandable that the measurements of the body and arms, made under these conditions could only indicate the proportions between different parts but not the exact size of each part of the living animal.

I give this cephalopod the name *Melanoteuthis lucens,* considering its coloration at capture and the exceptional presence of luminous organs.

Plate V depicts the entire animal from the ventral side. This strange grimacing mask has been featured according to Tinayre's water-color sketch made at the moment of capture and from my partial sketches of the preserved animal. The overall illustration is accurate but I have my reservations with regard to details such as the tip of the arms—they are filiform, which was not shown in M. Tinayre's water-color sketch and they have not been able to adequately resist the action of the formalin.

I have very carefully reconstructed this figure (Plate V) while studying the specimen. It was then colored by Miss Vesque based on the important water-color sketch made by M. Tinayre. Miss Vesque made this beautiful sketch with the preserved specimen always before her.

23 The body is short and continues, without subocular constriction, directly into the head and arms. The whole forms a gelatinous mass extending to the base of the arms, the lower half of which is buried within this jelly-like mass of tissue terminated by the thin umbrella. The umbrella-like interbrachial web thus consists of a thick wall in the lower half and is membranous only at its edge.

In view of the condition of this fine membrane, only shreds of which remain, it is impossible to estimate the extent and shape of the swimming membrane and its relation to the lateral membrane at the end of the arms.

I could only make out that the umbrella is indented between the two ventral arms and consequently does not form a continuous rim as it does in other *Cirroteuthis.*

The tip of each of the eight arms is shaped like a slender whip, 4 to 6 cm long. The tips may be longer in the live animal as I am not quite certain that these are intact.

Two large fins, 40 mm in length and 22 mm in maximum width, are inserted at the middle of the dorsal side of the mantle; the greatest width is close to the dorsal insertion. The fins are strong, fleshy along the rigid posterior edge with a thin and undulant anterior edge. They are dorsally separated by approximately

25 to 30 mm; their two lines of insertion are oblique and converge toward a point which almost corresponds to the mouth (Plate XV, Fig. 2).

Through the tissues of the fins, rendered transparent by the disappearance of the black pigment, is seen a strong flat muscular plate extending from the base to the tip of each fin. The muscle band bifurcates at the base: one branch is short and ascends, the other is long and descends; they are light brown or orange and seem to be inserted in the internal cartilage.

I cannot precisely describe this cartilage since I have not dissected the specimen. It is transparent and very protuberant dorsally at the lower part of the mantle. It has two branches rising toward the fins. This is all I could see through the skin.

Two oval spots are found in a cutaneous fold between the bases of these two fins. They are luminous organs and will be described later in the text.

On examining the ventral side of the animal, one is first struck by the enormous size of the mantle opening (Pl. XV, Fig. 3). It is usually small in the Cirroteuthidae. On the contrary, here it is a wide opening through which one can see the entire funnel and its folds, the free end of the rectum, the two renal openings like protuberant rosettes, and the tips of the two gills. The lateral commissures of this wide opening merge with the soft tissues surrounding the eye.

24 Above this large slit is the anterior opening of the funnel, a simple, round aperture embedded in the soft tissue that masks the entire organ.

The eyes are also entirely buried in the tissue infiltrated with water; only the circular opening of the brown pigmented eyelids are visible. Handling the specimen has made the ocular globes more visible; they are only slightly accentuated protuberances in the live, pigmented animal.

The dissection of the smaller of the two specimens provided interesting features of this eye. Its internal part is in the shape of a truncated cup with a flat base enclosing a very large spherical lens. The walls of this special cup are strongly pigmented with violet-scarlet, while the flat base is a lighter pinkish-lilac hue.

The rather short yellow optic nerve penetrates the center of the basal disk of the eye. It issues from a very large optic ganglion attached to the brain by a large commissure. This ganglion is completely separate from the brain. It is covered by a layer of very dark blackish-brown pigment over the entire convexity opposite the base of the eye; the half near the brain is light yellow and devoid of pigment. The optic nerve does not issue abruptly from the ganglion; it arises from a tapering point. Only the point at which the optic nerve emerges is it pigmented like the ganglion. The base of an unknown organ is attached on the ventral side above the pigmented border of the left eye. This organ, which has no analog in any other cephalopod, will be discussed later in the text. These anatomical features are shown in Fig. 6 of Pl. VI.

When I examined the larger of the two specimens, it measured 23 cm in length, excluding the terminal filaments of the arms which would add 3 to 6 cm.

The body is 56 mm wide at the eyes; the mantle opening is 43 mm. The anterior opening of the funnel and the two ventral corners of the mantle constitute the three vertices of an equilateral triangle. Each side is 30 mm.

Due to their bad condition, it is impossible to make accurate measurements for fear of breaking the arms; however, their approximate lengths are as follows:

1st dorsal	left	100 mm,	1st dorsal	right	104 mm
2nd	”	105	2nd	”	116 ”
3rd	”	102 ”	3rd	”	94 ”
4th ventral	”	110 ”	4th ventral	”	115 ”

25 As already mentioned, the greater part of these arms lies in the thickness of the hydrated umbrella. As such, only their free tips are visible from outside; they form only a small protuberance in the interior of the umbrella.

From the figures of the whole body and the specimen I studied, it appears that the arms have a larger free part, but since the umbrella between all the arms was torn, I could not make out the level at which they were attached to the web. They definitely did not have a free part as long as that shown in Fig. 1 (Pl. II).

An examination of the surface of the arms turned toward the interior of the umbrella shows the principal features of this genus *Melanoteuthis* (Pl. XV, Fig. 4).

First, the suckers are present only on the distal third of the arms. I have not found this in any other Cirroteuthidae. Consequently the surface of the first two-thirds of the arms from the mouth is *smooth and without suckers*; they only start at the level which seems to correspond to the free edge of the umbrella. This character is fundamental. The cirri are in two rows along the margins of the arms; they are short and papilliform close to the mouth and become longer and stronger from where the suckers begin. I have drawn a portion of an arm in the region 3 or 4 cm from the tip where the cirri are most strongly developed (Pl. XV, Fig. 5).

The cirri are conical and terminate in a papilliform point; the largest are 1 cm long.

Along the margins of each arm there is a narrow membrane which joins the umbrella; it narrows and disappears at the tip and is strongly pigmented (Pl. XV, Fig. 5).

Due to the poor preservation of the animal, I could not study the details of all the arms. Several peculiarities on a few of the better preserved ones are given below.

The first dorsal left arm which has a special basal organ to be described later in the text, is without suckers up to the 14th pair of cirri from the mouth; the suckers begin only at the 15th pair of cirri and there are only nine suckers in total which become progressively smaller; a few very small ones may possibly have been present at the base of the arm but have disappeared; the terminal tip is very long.

On the second left dorsal arm the suckers are present from the 9th pair of cirri; I have counted 21 suckers, the last few of which at the end of the arm are microscopic and very difficult to discern.

On the right ventral arm, there are 17 suckers; the last few are scarcely distinguishable. The first sucker is at the level of the sixth pair of cirri from the mouth.

On the third right arm, I counted 19 cirri; the first sucker is at the level of the seventh pair of cirri.

26 Between the two ventral arms, the web is indented so that these arms are more free than the others where the web is attached closer to their tips and is continued by the marginal membranes.

The suckers are small, round, widely open and globular; their cylindrical cavity is strongly folded; they are externally pigmented except for a white margin around the opening; the suckers are sessile although their base is constricted. Suckers on arms ranged from 9 to 21 in number, but those closest to the filiform tips are so small and so delicate that a few might have been missed on the least well-preserved arms. They alternate with the pairs of cirri. They were probably not present on the filiform part of the arm.

The mouth is broad, round, surrounded by a protuberant yellowish lip in the form of a ridge showing the beak.

An organ shaped like a long papilla (Pl. VI, Fig. 6) is located within the umbrella. It emerges from a pit in the pigmented skin near the mouth—between the base of the first and second left dorsal arms, closer to the first. This papilla is approximately 1 cm long is topped by a small white cylinder rolled into a spiral at its tip (Pl. XV, Fig. 6). The papilla is streaked, and pigmented brownish-violet; the terminal cylinder is colorless. Due to the transparency, a white thread is visible which follows the entire axis of the unpigmented part and terminates abruptly at its tip. It penetrates the papilla where it becomes curved.

I dissected this organ in the smaller specimen and discovered its principal relationships. However, sections will have to be made to discover the histological nature of this organ.

The base of the papilla is above the left eye, in direct relation to the pigmented cylinder enveloping the internal organs of this eye. A nerve strand penetrates the papilla and occupies the axis of it. This nerve emerges from the brain at the junction of the optic ganglion with the cerebral ganglion; it makes an S-shaped curve between its point of issue and its point of penetration into the papilla. The basal curve of the S is longer than its curve of penetration into the papilla. At its origin, this nerve is in contact with two others that ascend along the length of the pigmented peribuccal membrane; however, I could not follow the path. On the right, the special nerve is absent, the nerve fascicle thus contains only two nerves, whereas the left fascicle has three. This arrangement is drawn (Pl. VI, Fig. 8) from the dissected material.

No hypothesis can yet be formulated on the role of this asymmetric organ. I report its presence, while awaiting other specimens that could provide better information on its relation and function. At present, it can only be said that it is a sensory organ directly connected with the brain by a very important nerve that traverses its entire length, but it is impossible to state whether this sensory organ performs an olfactory function or some accessory function of reproduction.

27 It is observed that no other cephalopod has any similar organ particularly in the interior of the umbrella.

The funnel does not protrude from the skin. Its anterior opening is almost round, approximately 5 cm in diameter. It is wide open and through it can be seen a transverse, oval-shaped, highly developed, scarlet valve (Plate XV, Fig. 7). Above the valve the internal mucosa of the funnel is also red and has small folds converging toward a median pit between the valve and the funnel opening.

The posterior of the funnel has two lobes corresponding to the insertion of two retractor muscles. The entire enlarged part of this funnel is visible, due to the considerable depression in the marginal ridge of the mantle opening.

The most interesting peculiarity of this unusual cephalopod is the *dorsal* position of two large *luminous organs* between the bases of the fins.

The dorsal location of these luminous organs is *unique* among all cephalopods known to date; their homolog has not been found. This is also the *first time* that these organs have been detected in the octopods.

Each is an oval-shaped irridescent structure, placed between two lidfolds in the form of bridles. These are oblique in relation to the base of the fins to which they are very close. The tops of the two organs are 19 mm apart.

The two lidfolds resemble the eyelids of the human eye in shape and position; they open to a slightly convex, translucent, oval-shaped lens, 4 to 5 mm long and 3 mm wide, through which shines an iridescent strip, analogous to a carpet, flecked with green and orange. Numerous chromatophores are between the lens and the carpet; they do not seem to be at the same level. The shallowest chromatophores are brown while the deepest ones are red and orange. The lens is embedded in a very dark blackish-brown membrane. Due to the transparency of the neighboring hydrated tissues, a black nodule can be seen under each lens. It protrudes downward, is hard and corresponds to the light-producing part of the organ.

As far as can be judged from the arrangement of the muscular fibers, the lidfolds must be able to close over the lens, either to shade it from light or protect it.

It is quite remarkable that the pigment, which disappeared from almost every part of the body of this animal, still remained around these luminous organs.

On obtaining permission to remove one of these two organs I made some histological sections. Unfortunately, the animal had not been preserved well while it was fresh; therefore, the sections did not provide all the desired results. However, they provided interesting indications.

28 Figure 1, Pl. VI shows a cross section, magnified 30 times, which passes almost through the middle of the organ. The following features can be distinguished: 1—two ridge-like pigmented lidfolds which are visible on the right and left of the drawing; 2—a red fibrous cup; 3—granular mass embedded in the center of the fibrous cup; 4—thin transparent layer covering the fibrous core like a corneal plate; 5—a layer of chromatophores intercalated between the cornea and the entire free part of the organ; 6—a black pigmented fibrous layer enveloping the entire convex surface of the organ in contact with the ambient connective tissue.

Some supplementary information is necessary to understand this interesting organ.

The lidfolds are formed from large, loose connective tissue; longitudinal and transverse muscular bundles (yellow and red in Fig. 1, Pl. VI) are entwined in its fibers. At higher magnification the distribution of these fibers is seen to have an interesting peculiarity. These fibers are grouped into small bundles of threes or fours; the nuclei of their connective tissue sheaths are situated between them; this group is then enveloped in a black-pigmented connective tissue cover (Pl. VI, Fig. 2). Also there is a border on the lidfold like a black network containing small yellow bundles of muscle fibers. The surface of the lidfold is covered by an epithelium of a single row of flat cells with large nuclei. A row of large bundles with a pigmented sheath is immediately below the epithelium; a second row of finer bundles with one or two muscle fibers underlies the first row; finally, the center of the lidfold consists of less pigmented connective tissue with longitudinal muscle fibers. This lidfold arrangement and its pigmentation is extremely odd.

The lidfold pigment covers the entire mass of the luminous organ. It is a layer of connective tissue fibers in contact with the light-producing organ on which the black pigment is spread. The black fibers most closely in contact with the organ are tightly pressed and run parallel with each other (Pl. VI, Fig. 3); with increasing distance from the organ, they become no longer crowded, less regular and eventually without pigment. Thus, they are an integral part of the general cutaneous connective tissue.

This black conjunctive pigment is replaced by a layer of chromatophores of different colors in the free part of the luminous organ (Pl. VI, Fig. 1).

The general mass of the luminous organ consists of a cup of dense padded connective tissue (Pl. VI, Figs. 1 and 3). Parallel fibers are present at the surface of the convex part and at depth in the organ, making it denser. Similarly, the entire part surrounding the central concavity consists of dense tissue with closely packed concentric fibers. These fibers, mixed with some surface muscle fibers, spread out fan-like (Fig. 1) below and around the cornea at the surface. The bright red nuclei are distributed in the mass of the connective tissue network of this part of the organ.

29 The center of the organ consists of a tightly packed mass of reddish-orange cells. They have different sizes: the largest (Fig. 4) is approximately 0.0264 mm and the smallest approximately half that, i.e., 0.0129. Their nucleus is spherical or lenticular, with a granular content. These cells are closely packed in a compact mass; I could not ascertain more regarding their structure due to a fault in mounting the fresh tissues.

The cornea, if this name can be given to a transparent plate protecting the luminous organ, is a simple, thin layer of lamellar tissue in which it can be seen that the lamellae become dissociated in the deep part in contact with the subjacent layer of chromatophores. I saw traces of a surface epithelium but again, due to faulty mounting, I could not see its structure. This layer covers the central cellular core and the surface of the connective tissue cup with fan-like connective tissue fibers. When the lidfolds close, the inner side touches this cornea.

There is nothing significant to report about the subcorneal layer of chromatophores. The drawings with a view of these multicolored chromatophores through the transparent corneal layer are more informative than the sections that do not give any special indications on the structure of these small organs.

The central nucleus of granular cells can be considered the light-producing tissue in this organ. The light issues directly outward through the cornea, however, the concave layer of connective tissue is also illumined and the fan-like fibers must filter this light, perhaps transform it, for example, by polarizing it. These rays can exit only through the subcorneal surface because the entire convex surface is enveloped in a thick layer of black pigment preventing emission everywhere else.

The role of the tinted chromatophores can definitely modify the color of the light emitted depending on which are open or shut.

I have not been able to see the penetration point of the nerve fibers into the organ; the sections only show a large number of vessels mixed with pigmented connective tissue (Pl. VI, Fig. 1).

Since they are large and there are only two, these luminous organs must be extremely powerful and capable of singular effects owing to the role of the chromatophores.

The dissection of the smaller specimen has shown other less important peculiarities.

The brain, seen ventrally clearly shows the coalescence of the ganglia into two rings: pedal and visceral.
30 The optic ganglia are very large; some black pigment grains partially cover them as well as the visceral ganglia.

The alimentary canal is entirely violet colored. It is broad on its passage through the nerve centers: the bilobed stomach is small; the hepatic gland is strongly developed.

The renal glands hang like a yellowish bunch of grapes behind the stomach.

It must also be noted that the aorta and its two branches are completely covered by black pigmentary granulations.

To sum up, this cephalopod is one of the most interesting of all the forms described to date. Of all the octopods, it deviates most from the fundamental type of this division.

It is affiliated to the family Cirroteuthidae by its fins, umbrella, cirri and single row of suckers. However, special and important features prevent its affiliation to any of the existing known genera of this large family. The genus *Melanoteuthis*, established for it, is characterized by the absence of suckers on the proximal part of the arms, the presence of a whip at the extremity of each arm, the presence of an unknown organ at the base of the 1st left arm in the umbrella and the presence of a pair of dorsal luminous organs at the base of the fins. A valve is located within the funnel.

Taking into consideration the importance of these differential characteristics, Cirroteuthidae can be divided into two subfamilies: one without luminous organs and the other with them. The latter includes *Melanoteuthis lucens*.

Family II. PHILONEXIDAE d'Orbigny, 1838

Genus *Tremoctopus* Delle Chiaje, 1830

Tremoctopus violaceus Delle Chiaje, 1830

Expedition of 1904: Stn. 1768, depth from 0 to 300 m*. Slightly to the east of Fer Island (Canaries). Large (wide-mouthed) Richard net.

To this species, I assign, with strong reservations, a young octopod whose body length is 5.5 mm from the mouth to the posterior end. The two ventral and the two dorsal arms are approximately 4 mm long, the two lateral arms on each side are in the process of budding.

31

Family III. ALLOPOSIDAE Verrill, 1881

Genus *Alloposus* Verrill, 1880

Alloposus mollis Verrill, 1880

Expedition of 1910: Stn. 3031, surface.

A brachial crown and a portion of a separated arm, probably but not definitely, from the same animal, were caught at this station.

The crown contains only the middle part of the arms, the buccal portion and the entire terminal part disappeared. This large fragment preserved in sea water and formalin has become transparent; a portion of violet membrane lining the inner side of the web adheres to it. Some suckers are intact on these arms; others have become detached. The arms are still attached to each other by fragments of soft, transparent membrane.

*In summary table at the end of the text this depth is given as 0–3000 m—General Editor.

The membrane has the consistency and appearance of the texture of a large medusa. No comment can be made on the morphological order of these arms since their relation with the mouth could not be judged. The only possible information that can be provided is the length and circumference of the arms.

The first arm is 43 cm long.

The second is 72 cm long with a circumference of 27 cm.

The third is 84 cm long, 28 cm around the proximal section and 19 cm around the distal section.

The fourth is 14 cm long.

The fifth is 41 cm.

The sixth is 70 cm.

The seventh is 39 cm.

The eighth is 18 cm.

I did not weigh this fragment but estimate the total weight at approximately 20 kg; this represented only a small part of the animal which must have been enormous.

32 Family IV. OCTOPODIDAE d'Orbigny, 1838

Genus *Octopus* Lamarck, 1799

Octopus vulgaris Lamarck, 1799

1799. *Octopus vulgaris* Lamarck (**47**), I, p. 18

Expedition of 1902: Stn. 1340. Fayal (Azores). A large female bought in the market.

Expedition of 1903: Stn. 1497, depth 208 m. In front of St. Sebastian.

A specimen with a rather small body and very long arms resembling *Octopus macropus*.

Expedition of 1910: Stn. 3026, depth 85 m. Gorringe Bank. A specimen found in the stomach of a *Serranus atricauda* caught by hook and line.

Octopus arcticus Prosch, 1849
(Pl. VII, Figs. 4 and 5)

1849. *Octopus articus* Prosch (**71**), p. 53
1873. *Octopus bairdii* Verrill (**84**), vol. V, p. 5

Expedition of 1898: Stn. 922, depth 343 m. Close to the southern point of Norway (Cape Lindesnaes). Trawl.

One male, 13 to 14 cm long, with very prominent dorsal papillae. A water-color sketch made at capture shows its brownish-gray coloration. Stn. 960, depth 394 m. Between Norway and Bear Island. Trawl.

A. Young female 8 to 9 cm long, with the web completely reversed. Subocular papilla highly developed.

B. Male 11 to 12 cm long.

Expedition of 1899: Stn. 1052, depth 440 m. Coast of Norway. Trawl.

A. Young female approximately 6 cm long.

B. Young female approximately 4 cm long.

In my opinion there are no satisfactory figures of this animal, so I have reproduced photographs of two males caught at Stations 922 and 960. One photograph shows a dorsal view of the animal and the other gives details of the hectocotylus. It is pointless to add to its description since these photographs are adequately representative (Pl. VII, Figs. 4 and 5).

33 *Octopus* sp.?

Expedition of 1904: Stn. 1715, depth from 0 to 1,000 m. Large Richard net. Canaries.

A young specimen in poor condition. Total length—5 to 6 mm. The body apparently covered with small tubercles or bristles.

Genus *Eledone* Leach, 1817

Eledone cirrhosa (Lamarck) d'Orbigny, 1838

1799. *Octopus cirrhosus*, Lamarck (**47**), I, p. 21, Pl. I, Fig. 2
1814. *Ozoena aldrovandi* Rafinesque (**72**), p. 29
1838. *Eledone cirrhosus* d'Orbigny (**63**), p. 79, Pl. II
1843. *Eledone pennantii* MacGillivray (**17**), p. 31
1851. *Eledone aldrovandi*, Vérany (**83**), p. 12, Pl. II, III
1851. *Eledone genei* Vérany (**83**), p. 16, Pl. I
1886. *Eledone cirrosa*, Hoyle (**25**), p. 102
1901. *Moschites cirrosa*, Hoyle (**26**), No. 9

Expedition of 1900: Stn. 1086, depth 64 m. Southern coast of Ireland. One small specimen remarkable for the development of its villi. Station 1088, depth 20 m. Loch [*sic*] Juadh, west coast of Scotland. A specimen caught in a trammel net.

Expedition of 1903: Stn. 1535, depth 132 m. Anchored at the mouth of the Odet River [France]. One young specimen, very villous. Station 1540, depth 140 m. En route to Brest. One average-sized specimen, slightly villous. Beam trawl.

Expedition of 1908: Station 2792. Coast of Norway. Two good sized specimens.

Genus *Eledonella* Verrill, 1884

Eledonella diaphana Hoyle, 1886

1885. *Japetella diaphana*, Hoyle (**23**), vol. 1, p. 232
1886. *Eledonella diaphana* Hoyle (**25**), p. 107, Pl. IX, Figs. 3–6
1910. *Bolitaena diaphana*, Chun (**9**), p. 20

Expedition of 1901: Stn. 1182, depth 2,478 m. Cape Verde Islands. One specimen. Trawl.

34 Expedition of 1905: Stn. 2168, depth 0 to 2,000 m, Azores. Large Richard net.

Expedition of 1910: Stn. 3028, depth 0 to 5,100 m. Rapidly hauled in Bourée net, one specimen. Station 3039, depth 0 to 3,660 m. Rapidly hauled in Bourée net, one specimen.

Eledonella sp. ?
(Pl. VIII, Figs. 8, 9 and Pl. XV, Fig. 8)

Expedition of 1908: Stn. 2714, depth 0 to 1,400 m. Near Gibraltar. Large Richard net.

This young octopod, approximately 9 mm long, looked like a small transparent globe, with eight short arms projecting from it. The only opaque parts of the body were the two black eyes and a very oblique, fusiform mantle which was small in relation to the size of the animal. The eight brachial nerves, issuing from the brain and swelling out at each sucker in as many oval-shaped ganglia, can be seen exceptionally clearly. There are eight of these swellings per arm (Pl. XV, Fig. 8, *N.B.*).

The ventral arms are half as long as the dorsal, the others are intermediate.

The eyes are transversely elongated in the form of a wooden shoe, with a lens in the cavity. The surface of these organs is silvery (Pl. XV, Fig. 8, *O*).

The brain (C) and the optic ganglia (G. O.) are highly developed, each is almost as large as the other. The two pallial ganglia and the two nerves linking them to the brain can be clearly seen.

SUBORDER DECAPODA LEACH, 1818

Division A. MYOPSIDA d'Orbigny, 1845

Family I. SEPIOLINI Steenstrup, 1861

Genus *Heteroteuthis* Gray, 1849

Heteroteuthis dispar (Rüppell) Gray, 1849

1845. *Sepiola dispar,* Rüppell (fide Vérany) (**82**), p. 26
1839. *Rossia dispar*, Vérany (**82**), p. 63
1849. *Heteroteuthis dispar* Gray (**21**), p. 90

Expedition of 1905: Stn. 1965, depth 0 to 2175 m on a bottom of 2,225 m. Near Monaco. Large Richard net. A young individual, approximately 1 cm long from the mouth to the posterior tip of the body.

Genus *Rossia* Owen, 1834

Subgenus *Franklinia* Norman, 1890

Rossia caroli Joubin, 1902
(Pl. VIII, Figs. 1 to 4)

1902. *Rossia caroli* Joubin (**38**), p. 138
1902. *Rossia caroli*, Joubin (**39**), p. 80

Expedition of 1901: Stn. 1118, depth 1,098 m. Azores. One specimen caught in a trawl.

I gave a description of the first of the two specimens of this species in 1902, and I feel that I do not need to modify it to any extent. I reproduce it almost exactly here; the description agrees almost exactly with my later review.

This cephalopod is in an excellent state of preservation in 4% formalin.

The most striking feature is the high proportion of the width of the head in relation to that of the body, particularly because of the large protuberant eyes which touch almost on the median dorsal line. Thus the relatively small and conical body contrasts with the large head.

The skin seems to be absolutely smooth; but if the animal is allowed to dry slightly a few small low papillae become visible; they are only slightly protuberant and can therefore be easily overlooked.

The arms have only two rows of suckers indicating the assignment of this specimen to the subgenus *Franklinia.* They are on the strong arms.

Here are the dimensions of this specimen in millimeters:

Dimension	mm
Total length, not including the tentacles	112
Width of head, including eyes	43
Width of head at base of arms	25
Width of head at the neck	22
Length of tentacles	90
Length of tentacular club	20
Length of 1st arm (dorsal)	43
Length of 2nd arm	47
Length of 3rd arm	50
Length of 4th arm (ventral)	44
36 Length of body, dorsomedial	38
Length of body, ventromedial	31
Distance between the two free edges of the fins	59
Average distance between points of fin insertion	30
Width of mantle opening	38

We will now recapitulate the characteristics of the animal and describe them in some detail.

Head. It is almost entirely occupied by the eyes which meet near the dorsomedian line with a very small space between them. On the ventral side, there is a corresponding median depression between the two eyes, the furrow in which lies the top of the funnel.

The skin of the head appears at first glance to be entirely smooth and flat, both anteriorly and posteriorly, but under oblique lighting it is seen to be covered with minute granules, particularly above the eyes and at the base of the two dorsal arms.

The eyes have no eyelids; it is not clear whether there is a slightly protuberant, keeled ridge on the ventral side below the eyeball where it meets the cornea; the iris is slightly oval-shaped, almost round.

Arms. They are strong, without a swimming keel, linked by a short web between the dorsal and lateral arms which is absent between the ventral arms. The web only reaches the level of the third or fourth sucker on the lateral arms. The suckers are arranged in two rows along the entire length of all arms; they become progessively larger from the base to the middle of the arm and then progressively smaller toward the tip.

The *suckers* are spherical, slightly flattened, inserted on their peduncle by a point on their equator, in a notch on the meridian. The opening is very small, with a toothless horny ring which is slightly protuberant like a small bell-shaped mouth. The peduncle is slender, extremely short, with a very slightly protruding nipple on the apex. This peduncle is not very solid and can be easily detached.

Both dorsal arms are hectocotylized. Their suckers, in two rows, are almost uniform and are smaller than those of the other arms where the middle suckers are markedly larger than those of the proximal or distal parts. A marked keel is present all along the ventral edge of the arm; it disappears at the last suckers. It is internally furrowed by secondary slanting grooves that start at the intervals between the peduncles of the suckers. A similar keel is present along the dorsal margin but less clearly visible than on the ventral margin. The fleshy peduncles of the suckers of the hectocotylized arms are much longer than on the other arms.

37 *Tentacles.* They are almost cylindrical, rather large, and traversed along their entire length by a furrow which reaches the base of the club. The tentacles are longer than the body and head together.

The club is narrow and is simply a prolongation of the tip of the tentacle. A slight swimming keel occupies approximately one-third of the ventral edge. On the dorsal margin a very thin membrane borders the suckers along the entire length of their extent. These suckers are very small and their brown color contrasts with the white tentacle. They diminish in size almost imperceptibly from the base to the tip and are arranged in 6 to 7 rows. The suckers on the inner margin are no larger than those on the outer edge. Their large opening is in the form of a beaklike funnel. The bottom of the opening is surrounded by a dozen spongy teeth, less protuberant near the beak. Four or five rows of brown, horny tubercles like concentric pearls, are found on the widened part of the funnel, more numerous and smaller at the free border than near the teeth of the internal opening.

These suckers are covered within by the prolongation of the sheath which gives them a yellow cast. The laterally inserted peduncle is very slender.

The *funnel* has a rather narrow tube which does not reach the level of the bifurcation of the ventral arms; its base is very broad, with two very deep and broad adhesive grooves. The funnel organ consists of four well developed oval parts, two of which are dorsal and two ventral. The two dorsal parts are apically united and extend to the large valve, are well-formed and can completely close the tube of the funnel.

The mantle ridges, corresponding to the grooves of the funnel, are well-developed and stronger than in the other species of the genus *Rossia.*

The *body* is conical, very tapered, short and posteriorly pointed. Dorsally, the skin is ornamented with very small papillae, scarcely noticeable like those on the head. The dorsal mantle margin has a very marked projection in the form of an obtuse angle lying on the neck cartilage but without a protuberant tubercle at this tip. The nuchal cartilage is broad, oval-shaped, has free edges and a strong median furrow where the tip of the gladius lies.

The ventral mantle margin shows a slight indentation under the funnel, and two more equally attenuated indentations are below the eyes.

The *fins* are small, semicircular and inserted much closer to the dorsal mantle margin than to the tip of the body. They are inserted obliquely at the extreme edges so that they are far apart; they have an apical indentation but not a basal one.

A second specimen caught at Cap d'Ail near Monaco in 1910 (EIDER, Stn. 01276) is almost the same size as the first. It is whole and in excellent condition except that the posterior part of the mantle has been slightly flattened during storage in a flask. The hardening effect of formalin has strengthened this slight deformation. This has slightly changed the appearance of the animal and makes it look different from the first specimen (Pl. VIII, Figs. 3 and 4).

38 This female specimen is about 100 mm long, including the tentacles, and 70 mm between the posterior end of the body and the tips of the arms. The latter measurement is slightly approximate due to the flattening already mentioned. I have sketched this animal from the ventral side and considerably magnified it.

I now must examine the affiliation of this cephalopod with other species of the same genus.

It is known that forms belonging to *Rossia* are divided into three subgenera. If only the left dorsal arm is hectocotylized, then the forms belong to *Semirossia.* If both dorsal arms are hectocotylized with four rows of suckers then these belong to the subgenus *Rossia*, or if they only have two rows, they belong to the subgenus *Franklinia.* Our species belongs to this last subgenus.

The subgenus *Franklinia* includes only three species: *Rossia glaucopis* Lovén, *R. subloevis* Verrill and *R. megaptera* Verrill. The latter two species can be immediately eliminated since, even at first glance, they differ so noticeably from our species that further examination is pointless. Our species can only be compared to *R. glaucopis.*

R. glaucopis is an exclusively polar species; the most southern point at which this species is found is the Lofoten Islands. It has a large, thick, posteriorly obtuse body. The short tentacles do not extend beyond the middle of the mantle. The slightly prominent eyes have thick, wrinkled eyelids. The arms are short and united by a thick web. The dorsal cutaneous papillae are protuberant and highly developed, particularly those on the tip of the dorsal mantle angle. The dorsal margin of the mantle is straight.

On the contrary, in *R. caroli* the tentacles are long, the body small and posteriorly pointed. The dorsal margin of the mantle has a broad angle, the skin is almost smooth and the brachial web is almost absent. It lives in warm seas (Azores, Mediterranean).

Rossia glaucopis Lovén, 1845
(Pl. VIII, Figs. 5 and 6)

Expedition of 1899: Stn. 1052, depth 440 m. Coasts of Norway. Trawl.

Only one specimen, a female in very good condition, approximately 5 m long including the tentacles. Spermatophores still adhere to the area above the left eye. A water-color sketch of the live animal was first made by Mr. W. Smith. It has an overall grayish-blue tint, more accentuated on the head, with green above the eyes, the head and on the dorsal arms.

I photographed the animal preserved in formalin, shown in Figs. 5 and 6 of Pl. VIII.

39 Subgenus *Rossia* Owen, 1834

Rossia macrosoma (Delle Chiaje) d'Orbigny, 1839

1828. *Sepiola macrosoma,* Delle Chiaje (**10**), p. 71
1839. *Rossia macrosoma* d'Orbigny, (**62**) p. 245

Expedition of 1903: Stn. 1455, depth 358 m. Open waters of the mouth of the Gironde. Beam trawl. One male specimen of medium size.

Stn. 1497, depth 208 m. In front of St. Sebastian. Beam trawl. A very large female specimen with body length approximately 18 cm without the arms.

Rossia palpebrosa Owen, 1835

1835. *Rossia palpebrosa* Owen, (**65**), p. 93

Expedition of 1908: Stn. 2786, depth 17–20 m. Thamshavn (Norway). One adult female specimen caught in a trammel net.

This species is possibly the Arctic form of *R. macrosoma,* which is more southern. It strongly resembles it.

Family II. SEPIARII Steenstrup, 1861

Genus *Spirula* Lamarck, 1801

Spirula sp.
(Pl. IX, Figs. 1 to 5; Pl. XVI, Figs. 1 to 19)

Expedition of 1901: Stn. 1222. On the beach of Tarrafal, Cape Verde. A large number of empty shells.

Expedition of 1904: Stn. 1763. On the beach of Las Palmas (Grand Canaries). A large number of shells, a few of which I refer to. Several shells, still have soft fragments adhering. Stn. 1768, depth from 0 to 3,000 m. Large Richard net. Southern coast of Hierro Island. Canaries.

The first two stations provided many empty *Spirula* shells; their only interest is to record their location.

40 However, the specimen from station 1768 is extremely interesting and important to the study. In fact, this is the first known specimen of an original *Spirula.* I think it is necessary to reproduce here the description of this animal I published in the *Bulletin de l'Institut Océanographique* (**42**).

Among the animals which have long excited the curiosity of naturalists because of their extreme rarity and their great morphological interest, *Spirula* is first among mollusks; this distinction might even place it first among invertebrates.

Empty *Spirula* shells are abundant in most collections, but there are only a few whole specimens in good condition from which the anatomy of the animals can be studied. Among these, the two specimens magnificently described by Pelseneer (27a) and the live individual observed by Chun are noteworthy; only an external representation of the latter has been published.

To date, no one has seen the egg, embryo or the young hatchling of *Spirula*. All that we know reduces to the probability that, judging from the size of the ovarian follicle of the adult, the egg must be rather large. Probably the size of the adult has the same relation that the egg size of the cuttlefish has to an adult cuttlefish. This concept, as yet not confirmed, is the basis of our knowledge of the *Spirula* apart from the main structure of the adult.

An interesting report has enabled me to throw some light on the unknown evolution of the *Spirula*.

During the cruise made by HRH, the Prince of Monaco, in 1904 and in which I participated, a catch was made on August 17 at 27°43′ N and 18°28′ W, by the large bathypelagic Richard net. This station No. 1,768 is located southeast of the Canary Islands close to the southern coast of Isla Hierro. The net was lowered to 3,000 m. The small cephalopods were brought to me after the plankton was removed from the net. Among the cephalopods I found a small *Spirula*, a little more than 5 mm in total length. The specimen was entirely intact but its long storage in alcohol had somewhat deformed it. The retracted skin had developed numerous wrinkles and a part of the epidermis had been chaffed off, particularly on the head and arms.

I photographed this animal from different views and reproduce here the best pictures to give an adequate idea of the report I had made (Pl. IX, Figs. 1 to 5).

The total length of the animal, including the very short arms, is 5.35 mm. Its width at the middle of the mantle is 3.10 mm, and 2.75 mm at the middle of the shell.

The following features could be identified from an external examination of this young cephalopod.

41 The body is highly contracted and the head is largely withdrawn into the mantle cavity with only the arms, the beak, and the surrounding lip protruding from the opening. The margin of the mantle had to be pushed back with a needle to see the eyes. This general contraction imparted a squat appearance to the animal which would probably not be so accentuated in its living state. This contraction has had an uneven effect because the anterior part of the body is spongy, whereas the posterior part around the shell must have changed very little.

The anterior portion of the mantle is almost cylindrical while the posterior part with the fins and shell is laterally flattened (Pl. IX, Fig. 1).

The head is scarcely visible because the arms surrounding it project very little, being reduced to small stumps applied against the mandibular bulb. At the time of immersion in alcohol, the head was turned slightly askew from the normal sagittal plane by a light torsion so that its axis no longer corresponded to that of the body. The center of the head has fairly strong black jaws projecting from a thick circular lip slightly indented at the sides forming a striated ridge around the buccal opening. The radula can be seen between the jaws.

I could find no trace of any stalk of a yolk sac between the arms, from which I concluded that it had been resorbed and that the animal had no vestiges of it at hatching.

The eight arms (unequal stumps) are distinctly visible, but this is not the case for the tentacles and I have not confirmed their presence or absence. I cannot say whether the very small tubercles, seen in their place (Pl. XVI, Fig. 2), constitute the initial stage of tentacle formation, or, to the contrary are their last remnants after being detached. Perhaps the animal defended itself at the time of capture and lost its tentacles in the scuffle. An indistinct sign of mutilation is visible around these stumps, but I do not know whether this appearance is due to the peeling of the skin during its storage in alcohol or to the tearing off of the tentacles. I do not know whether they could have been retracted into a supraocular pocket, which would be surprising for such small organs. The difficulty of making a definite statement is understandable considering that the organs are only 1/10th to 2/5th of a millimeter in size and in poor condition.

However, 4 brachial stumps are clearly distinguishable on each side and a fifth is very indistinct. The two pairs of dorsal arms are the largest and distinctly extend beyond the buccal bulb; the third pair is almost the same length as the edge of this bulb; the fourth pair is very small, and an empty space separates it from the third pair. The three large pairs of arms have about 20 small suckers, each with a slender stalk, a dark brown horny ring on the oval surface which does not descend to the base of the arm. There are only five
42 or six suckers on the fourth pair of arms.

In this state of contraction, the eyes are not visible on the surface of the head protruding from the margin of the mantle. The mantle border had to be moved aside with a needle to see them. Even then, they are difficult to discern as they are very slightly protuberant, flat, small and seem to be buried in the muscular mass of the base of the arms. Moreover, they are colorless, a fact that differentiates them from the eyes of young cephalopods which generally are large and highly pigmented even before hatching. These pigments are normally resistant to the action of alcohol; in this case, if pigmentation had been present it was not resistant. I rather believe that, in view of the underdevelopment of eyes, pigmentation was absent or only slightly present. Thus, the eyes of this young *Spirula* are strongly underdeveloped in comparison with the general development of the other organs and with the development of the same organs in other Cephalopoda. The eyelid surrounding and enveloping the eye in the adult here is reduced to a slight fold of skin around the small, flat, very slightly raised lens representing the cornea.

This reduced eye contrasts singularly with the highly developed, protruding, conical eye depicted by Chun for the live adult *Spirula.* This clearly indicates therefore a *telescopic* aspect. However, nothing is yet known of its histological structure. This difference between the telescopic shape of the adult eye and the rudimentary flattened, not protruding shape of the eye of the young, closely coincides with Brauer's discovery in young bathypelagic fish, which have a hemispherical eye that later transforms into a very elongated, cylindrical eye. The formation of the retina, which is atrophied at some places and hypertrophied only at the base of the eye, is very interesting. I could not confirm whether these two structures extend further, but they are externally visible in young and adult *Spirula.*

Another noteworthy point is the relatively enormous development of the funnel. In order to see it, one has to push back the edge of the mantle on the ventral median line, then from the dorsal side and finally the lateral sides (Figs. 5 and 7). Thus one can see that the central tube is a strong, thick, muscular, squat cylinder, which descends very deeply into the mantle cavity. The end of this tube is lodged in a very distinct depression at the base of the ventral arms between the two eyes. The lateral pouch valves on each side of the funnel are highly developed and meet at the median dorsal line (Pl. XVI, Fig. 5) of the nape where only a narrow triangular patch separates them. This entire propulsion apparatus is very powerful in comparison to the small size of the young *Spirula*; it occupies the greater part of the mantle cavity. This young cephalopod thus must be a very active swimmer.

The mantle covering the entire visceral cavity is very wrinkled by the action of the alcohol. It must
43 be smooth and semi-transparent in the living animal. Almost its entire length is sprinkled with chromatophores both over the shell as well as on the part corresponding to the viscera (Pl. IX, Figs. 1 to 5). I could not see whether these chromatophores, which had not been fixed, presented any characteristic features.

In the adult, the mantle margin is provided with three very marked projections or corners, two ventral around the funnel and one dorsal. In the young, it is almost smooth and continuous except for a very slight depression on the dorsomedian line and two other lateral depressions at the position of the eyes: however, the latter are so slight that they may simply be due to the action of the alcohol. The mantle contains no trace of the feature that later forms the two windows exposing the shell dorsally and ventrally in the adult. Similarly, there is no trace of a suture which indicates that the shell is first external and then covered by the two lobes of the mantle which converge posteriorly and then unite. The mantle is smooth, continuous, scattered with chromatophores even where this suture should be located, as on all the rest of its length, and nothing indicates that it is abnormal. Based on these facts, it can be assumed that the two windows to expose the adult shell are produced much later by a wearing away of the mantle, which becomes thin due to the two protuberances of the shell which have become large and prominent. The examination of the *Spirula* first studied and so meticulously described by Pelseneer, does not leave me in any doubt with regard to this reasoning. Naturally, I cannot precisely say when the mantle is worn off; the only thing I can confirm is that nothing in the young individual I have studied indicates that this feature would appear later.

Moreover, there is no trace at the back of the shell of the unpaired, median, posterior sucker so characteristic of the adult *Spirula* (Pl. IX, Figs. 1 to 5). Pelseneer assumed that it is produced at the meeting point of the two mantle lobes enclosing the sides of the shell. This is not the case since these two lobes do not exist and there is absolutely no trace of them in the part of the mantle covering the shell where this unique sucker will later appear. A similar example of this sucker is not found in any other cephalopod. Therefore, a later formation of this organ must be sought for as it appears it does not function as a sucker. It probably appears much later, possibly at the commencement of genital activity and is related either to mating or egg laying. Its appearance, then, would coincide with the displacement of the fins, consecutive, as has been stated, to the development of the genital glands. One could also think it produced light which, however, does not exclude its affiliation with the genital function. In any case this organ is totally absent in the recently hatched young.

44 The fins are rounded, projecting and attached by a short muscular peduncle, corresponding to approximately one-fourth of their periphery at the mantle surface. They clearly resemble two small lateral ears. They are inserted almost at the center of the spire of the shell and are thus separated from each other by the thickness of the shell plus the mantle covering. An apt expression would be "one cannot see the other". This position completely changes in the adult when the fins are no longer lateral but terminal; they are no longer separated by the shell but are behind it. Juxtaposed, they can "entirely see each other," nothing is between them but an unpaired, low groove which prevents their bases from becoming confused.

The result of this placement is that, in the young *Spirula*, the fins have a provisional position destined to change considerably in the adult. They slide from the sides toward the posterior of the body at a time, as just mentioned, which must correspond to the development of the genital organs. They are forced back by the thrust of the organs that modify the position of the original visceral cavity. They occupy two lateral pockets on each side of the shell. These are absent in the young individual. These two additions to the visceral cavity contain the genital, digestive and renal organs, enveloping sides of the shell hiding it, and change the flat body of the young individual to the cylindrical body of the adult. The overall physiognomy greatly changes at this time, and after the formation of the telescopic eyes, these changes must coincide with the development of fins and the appearance of the terminal sucker.

I summarize in two diagrams of the structure of the *Spirula* in the young and adult stages (Pl. XVI, Figs. 6 and 7). I constructed the former from the young *Spirula* which was the object of this study, and the latter from the beautiful anatomical drawings by Pelseneer included in the outline of the live adult *Spirula* studied by Chun. In view of the above, lengthy explanations are not necessary to understand the overall arrangement. The noteworthy point is the very small visceral mass above the shell in the young individual as compared with the very large mass of similar viscera to the right and left of the shell in the adult. The important modifications that arise when this secondary arrangement develops will be discussed later in the text.

Only the shell now remains to be discussed. From the preceding comments it is clear that it is entirely subcutaneous. Therefore, to see it I would have had to cut the mantle enveloping it. I did not wish to do this because this *Spirula* would have to be dissected and I wanted to preserve it intact. Thus, all I could see of the shell was what I could discern through the skin and from the external projecting parts (Pl. XVI, Figs. 8 and 9). There are six chambers in all and I have noted their precise positions. In one of the figures

45 I have shown the shell in the exact position it occupies in the animal; this shows the position of the initial bulge when the shell has six chambers, i.e., almost at the time of hatching. I have added the shell siphuncule after Pelseneer, because what I could observe under the conditions was too vague for a detailed drawing. In order to check what the shell was like at this stage I detached the central part of a shell of an adult *Spirula.* Only the six initial chambers were present which gives an exact idea what the shell of a young *Spirula* is like.

I have repeatedly referred to the excellent works of Pelseneer throughout this paper. It is necessary to again refer to it.

In April 1895, Pelseneer (27b) published "Observations on *Spirula*" in the *Bulletin Scientifique de France et de la Belgique* in which he describes the anatomy of two adult *Spirula.* It is a major memoir full of facts and interesting deductions which could serve as a model, and which in any case establishes our knowledge on most points regarding the anatomy and morphology of this group of cephalopods. Searching to deduce the evolution of *Spirula* from particular resemblances to *Nautilus* and similarities with other families, he propounded a theory based wholly on assumptions about the initial stages of shell formation. Since neither Pelseneer nor anyone else had ever seen the embryo of *Spirula,* this theory was only an assumption, as admitted by the author. The discovery of the young *Spirula* has now negated Pelseneer's theory.

Since this very ingenious theory is very interesting, I thought it necessary to reproduce the three drawings of the author and cite the passage describing them (Pl. XVI, Figs. 10, 11, 12).

"It must thus be imagined that during development, the free edges of the mantle reflect over the shell as in many mollusks (particularly on the dorsal side of the shell of *Nautilus*), and the shell is thereby gradually covered (Fig. E). The right and left sides of the mantle evidently extend to the aboral end, the prolongations that unite on the median line creating the terminal disk are probably only slight extensions initially (Fig. F). It should not be assumed that the shell would have been completely covered very early and that the dorsal and ventral opening of the shell cavity would have been formed later from pressure caused by the shell on the thin regions of the mantle envelope. This pressure could be the only possible cause for the mantle to remain partially open allowing the slits to remain open. The development of the shell sac of the other cephalopods (*Sepia, Loligo*) is essentially the same except that the edges do not join only at one point but along their entire length."

I cannot accept Pelseneer's theory because it is marred by the fact that the young *Spirula* with six chambers has an entirely subcutaneous shell. Thus, the shell is not covered by the two bridles over the two openings. Contrary to Pelseneer's views I think that, as in other cephalopods, particularly *Sepia* and *Loligo,* a small shell invagination develops early in the middle of the dorsum of the mantle, which is quickly closed, creating a subcutaneous shell from the first moment. It remains in this state during the embryonic and juvenile stages and I think it is only much later, perhaps at the descent of the visceral sac compressing the fins, that the shell pierces the skin over it which has been stretched thin and gradually worn out. I believe the two openings appear secondarily when the shell has grown considerably and pierces the thinned skin over it. Sometimes the edge of the opening is uneven, as clearly shown by Pelseneer for one of his specimens, and the method of perforation of this opening is clearly apparent; sometimes the edge of this opening is smooth, having healed over or worn even at a later time, in a manner impossible to pinpoint. Our two theories thus differ on the fundamental point, viz., Pelseneer thought the shell of the *Spirula* is *initially external* and later *partially subcutaneous* (except for the two openings), whereas my view is that the shell is *initially internal*, formed in the mantle shell sac that arises from an invagination, and becomes *partially external* only when the two openings, which partially expose its last whorl, allow it to be visible through a later wearing away of the skin, by rupture or other means.

My theory has an advantage over that of Pelseneer's by virtue of being based on an accurate fact. It is a valuable landmark but admittedly insufficient for establishing a certainty.

Like Pelseneer, I have given a series of explanatory diagrams which are sufficient to allow one to see the difference between the two theories without need for further elaborations. The last two figures of the series show the stage with six chambers that I depicted earlier and can be intercalated here. Finally the adult form can be added where the shell (seen in profile) protrudes through the skin on the dorsal and ventral median lines.

It is now hoped that the new Richard fishing net will have been used during the same season south of Hierro Island. I think we will then stand greater chance of catching more young if the net is towed very slowly around 3,000 m. It would be equally important to use a trawl in this region to catch egg laying specimens where they are fixed to the bottom. If the bottom is rocky (which is possible in view of the entirely volcanic nature of the region) the dredge can be used.

47 *Spirula* sp.

(Pl. XI, Figs. 6 to 9)

Expedition of 1908: Stn. 2688, depth from 0 to 2,025 m. Balearic Islands. Large Richard net. Two specimens. Stn. 2704, depth fron 0 to 1,665 m. Between the Balearic Islands and Gibraltar. Large Richard net. One specimen.

Of all the cephalopods covered in this volume, those that have given me the greatest difficulty are the specimens described here under the name of *Spirula.* It is after great hesitation and after considerable correspondence with Dr. Pfeffer, that I have finally placed them in this genus. Dr. Pfeffer had occasion to see similar specimens possessed by one of his colleagues and this determined his conclusion.

Whatever the case may be, these unique embryos differ so much from those known to date among young cephalopods that their study would be of greatest importance. Unfortunately they were not fixed as soon as they were captured; they were mixed in a large quantity of plankton from which they were sorted out some years later.

The most important characteristic of these unique embryos, which are only 3.5 mm long, is the presence of a pair of small, triangular, *distinctly ventral* fins inserted on the mantle membrane, well below the edge of the muscular sheath enclosing the upper part of the mantle. These two white, opaque fins are on a thin transparent membrane covering the lower third of the body.

A very thin chitinous membrane envelops the entire embryo. It is so thin that its presence can be detected only by its slightly iridescent reflection under oblique light. Through this membrane one can distinguish a second sac, also transparent with granular walls, continuous with the inner side of the mantle muscular sheath. This second sac contains a large brown, bilobed, pear-shaped mass occupying the largest part of the visceral cavity. This is the vitellus which, maybe, is already in the process of forming the hepato-pancreatic gland.

The mantle muscular sheath is whole on the ventral side, but is split on the dorsal side. Its two halves are united by a very small bridle on the mantle edge. This sheath forms a double hump, dorsal and ventral, with a small ventral keel. The closing on the dorsum probably confines a ridge where the rudiment of the embryonal shell forms. The lower chitinous sheath thus envelops perhaps the mass of viscera in the granular sac. The mantle edge is wavy, strongly raised on the dorsal and ventral median lines, and cut away on the sides.

48 The head protruding from the mantle is almost whole. Five pairs of arms, large eyes and the largest part of the funnel can be distinguished here.

Only four pairs of arms are seen on one of the slightly younger embryos. Pairs 1 (dorsal), 2 and 4 (lateral) are well developed; pairs 3 and 5 are at the stage of rudimentary stumps. Some suckers without a horny ring can be distinguished on the most strongly developed arms. It is difficult to count them because the most distal ones are barely visible and some are probably broken off at the promixal part of the arm. Three or four pairs can be seen on arm 1, and five or six pairs on arm 4.

The eyes are large, telescopically shaped; their cornea is highly protuberant and the lens is very much smaller. On each side of the funnel, a white plate can be seen on the outer face of the body of the eye. It is probably the rudiment of a luminous organ.

The funnel is highly developed; it extends to the level of the lens. Its upper opening is a transverse slit with an upper and lower lip at right angles to each other. The lateral pouch valves of this funnel extend to the dorsomedial line; its anterior membrane is highly developed and supported by two strong and obliquely directed pillars.

Genus *Sepia* Linné, 1766

Sepia elegans d'Orbigny, 1829

1836. *Sepia elegans* d'Orbigny (**62**), p. 280

Expedition of 1903: Stn. 1475, depth 1409 m. Arcachon anchorage. Beam trawl. One male and three females. Stn. 1535, depth 132 m. Odet anchorage. Beam trawl. One pregnant adult female. Stn. 1540, depth 140 m. En route to Brest. Beam trawl. One small specimen in bad condition.

Expedition of 1905: Stn. 1994, depth 141 m. Near Monaco. Large beam trawl. Two adult females.

Sepia orbignyana Férussac, 1826

1826. *Sepia orbignyana* Férussac and d'Orbigny (**61**), p. 156

Expedition of 1903: Stn. 1475, depth 140 m. Arcachon anchorage. Beam trawl. One adult male.

Expedition of 1905: Stn. 1994, depth 141 m. Open waters off Monaco. Large beam trawl. One male and two adult females. Four young.

49 Division B. OEGOPSIDA d'Orbigny, 1845

Family I. ONYCHOTEUTHIDAE G. Pfeffer, 1900

Genus *Onychoteuthis* Lichtenstein, 1817

Onychoteuthis banksi (Leach) Férussac, 1817

1817. *Loligo banksi* (Leach) (**49**) Férussac (**12**), p. 332

Expedition of 1904: Stn. 1874, depth from 0 to 2,000 m over a sea bottom of 3,800 m. Large Richard net.

One young embryo approximately 6 mm in length in the stage represented by G. Pfeffer (**70**, Pl. 6, Figs. 6 and 7).

Expedition of 1905: Stn. 2130, depth from 0 to 3,000 m. Large Richard net.

One young specimen 4 to 5 mm corresponding to Fig. 24, Pl. 3 of the memoir of G. Pfeffer (**70**).

Expedition of 1908: Stn. 2695, depth from 0 to 2595 m. Large Richard net.

Two very young embryos approximately 3 mm in length, a stage corresponding to that which G. Pfeffer has shown in Figs. 4 and 5, Plate 6 (**70**).

Genus *Chaunoteuthis* Appellöf, 1890

Chaunoteuthis mollis Appellöf, 1890

1890. *Chaunoteuthis mollis* Appellöf (**2**), p. 29

Expedition of 1902: Stn. 1307, surface.

An adult specimen was captured. It was unfortunately in bad condition; the deteriorated head was without eyes and most of the neck folds. Only the stalks of the tentacles, each measuring 1 cm, were intact.

The body and arms are in good condition; with preservation in formalin their color is marbled-brown. The soft and almost gelatinous consistency, which led Appellöf to name this species *mollis*, has remained.

50 The total length of the animal from the posterior end of the body to the tip of the arms is 33.5 cm. The body measured from the funnel groove to the tip is 165 mm. The fins span 168 mm from tip to tip. The body is slightly more inflated above the fins than shown in the specimen illustrated by Pfeffer; its diameter is 67 mm at this point.

The arms measured from the middle of the mouth are: first arm (dorsal) 54 mm; second (lateral) 68 mm; third (lateral) 73 mm; fourth (ventral) 74 mm.

The buccal membrane has all the membranous attachments indicated by the authors, but its labial margin is slightly more dentate. There is a large valve in the funnel. The funnel locking cartilage is 26.5 mm long.

The fins are very broad and very long; the distance from the tip of the body to their apex is 124 mm, which is one-fifth more than in the specimen in the Museum of Upsala, cited by Pfeffer, which is only 100 mm for an almost equal body length; the width measured from the middle of the dorsum, is about 80 mm instead of 70 mm for the same specimen.

This interesting specimen is the first of this genus which is represented in the Museum of Monaco. A water color sketch of the animal was made by M. Borrel at its capture.

Family II. ENOPLOTEUTHIDAE G. Pfeffer, 1900

Genus *Abraliopsis* Joubin, 1896

Abraliopsis pfefferi Joubin, 1896
(Pl. XII, Figs. 1 and 2)

1896. *Abraliopsis pfefferi* Joubin (**34**), vol. 5, No. 1
1912. *Abralia* (*Abralia*) *pfefferi*, Joubin, G. Pfeffer (**70**), p. 156

Expedition of 1910: Stn. 3039, depth from 0 to 3,660 m. Bouree net drawn rapidly.

A young 17 mm long individual. I have shown it in Pl. XII, Figs. 1 and 2 because of the clarity with which it shows the position of the luminous organs which slightly protrude on the ventral side and can be differentiated from the chromatophores among which they are mingled. This animal shows all the characteristics of the adult except for its many fewer luminous organs.

Abraliopsis sp.
(Pl. XII, Figs. 3, 4, 5)

Expedition of 1904: Stn. 1834, depth from 0 to 1,000 m. Large Richard net. To the south of St. Miguel, Azores.

51 Approximately 7 mm long juvenile in too poor a condition for the species to be determined.

Expedition of 1909: Stn. 3902, depth from 0 to 1,800 m. Large Richard net.

Three juveniles—5.5, 6 and 11.5 mm.

I have shown photographs of these three embryos as they furnish three stages of larval development which probably result in the juvenile form of *Abraliopsis pfefferi* described in the previous chapter. However, two or three other intermediate stages would be necessary to confirm the identification. A striking feature of the youngest individual is the circular arrangement of the initial chromatophores; the luminous organs are not yet formed. The fins are typical of this form.

Genus *Thelidioteuthis* Pfeffer, 1900

Thelidioteuthis alessandrinii (Vérany), 1851
(Pl. XII, Figs. 6 and 7, Pl. XIII, Figs. 7 and 8)

1851. *Loligo alessandrinii*, Vérany (**83**), p. 99
1857. *Enoploteuthis polyonyx* Troschel (**80**), p. 41
1881. *Onychia caribaea* Steenstrup (**78**), p. 96
1881. *Abralia megalops* Verrill (**86**), p. 364
1884. *Enoploteuthis pallida* G. Pfeffer (**68**), p. 18
1889. *Calliteuthis alessandrinii*, Appellöf (**1**), p. 449
1900. *Thelidioteuthis polyonyx*, G. Pfeffer (**69**)
1851–1910. *Thelidioteuthis alessandrinii* Vérany (**83**), (Chun **9**)

The specimens of this species of cephalopod are all juveniles caught in a plankton net. The preceding abridged synonymy indicates how these animals are still poorly known and how much their classification is subject to variation. It is possible that we have not yet reached the end of our difficulties because the study of small planktonic cephalopods has scarcely begun.

Expedition of 1904: Stn. 1794, depth from 0 to 3,000 m. Large Richard net (Pl. XIII, Figs. 7 and 8).

A young, approximately 8 mm long, which corresponds to the stage of development illustrated by Chun (**9**, Pl. VII, Fig. 14). The ventral luminous organs of the mantle are not yet visible; the suckers of the tentacles are slightly larger than indicated by Chun.

Expedition of 1908: Stn. 2695, depth from 0 to 2,595 m. Large Richard net (Pl. XII, Figs. 6 and 7).

This young cephalopod is approximately 10 mm in length. Since it is slightly less advanced than the
52 individual illustrated by Chun (**9**, Pl. VII, Fig. 14), I consider it necessary to show a photograph of it (Pl. XII, Figs. 6 and 7). The head is large and the eyes very protuberant. The chromatophores are numerous on the ventral side of the mantle, and more sparse on the dorsum. The chromatophores on the arms are distinguished by their arrangement in a longitudinal series. The tentacles are highly developed and the suckers grouped at their tips are very protuberant. The funnel is large, curved toward the base; it projects from a well-marked notch in the mantle margin.

Expedition of 1909: Stn. 2916, depth 0 to 1,000 m. Large Richard net.

This is a young, almost adult specimen, measuring 13 mm from the mouth to the posterior end, 20 mm including the tentacles. It differs only in that the fins have not yet reached normal proportions. The chromatophores are large and numerous; the suckers are well-developed and all present.

Genus *Pterygioteuthis* H. Fischer, 1896

Pterygioteuthis giardi H. Fischer, 1896

1896. *Pterygioteuthis giardi* H. Fischer (**41**), p. 205

[For comprehensive synonymy see Pfeffer, (**70**, 1912, p. 204 and 777)].

Expedition of 1910: Stn. 3045, depth from 0 to 1,400 m. Bouree net drawn rapidly. Two specimens.

The smaller of the two specimens is broken in two, and the other is intact. These are two young females.

The complete specimen is 47 mm long, including the tentacles; the mantle is 22 mm measured along the ventromedial line and the tentacle is 18 mm measured from the mouth. The fins together have a width of 15 mm.

The broken specimen is only 16 mm long (as against 22 mm) along the ventromedial line.

A large cluster of spermatophores inserted on the skin of the mantle cavity can be seen close to the dorsomedial line on the broken specimen.

Two fairly large luminous organs are visible around the anus in the broken specimen. Those surrounding the eyes have been lost due to the ruptured skin; it is quite possible that they may be absent.

The tentacles are slightly swollen below the club region and, due to its transparency, a white, probably luminous nodule is visible. There are no hooks on the tentacular club; four rows of very small suckers are visible.

53 These cephalopods resemble the illustrations given by Chun (Pl. XII) for the male; these specimens are without the luminous organs which are located on the mantle skin in the male. This gives the specimens a strong resemblance to the illustrations of *Pyroteuthis margaritifera* published by G. Pfeffer (**70**, Pl. XIX, Figs. 17 and 18). However, the examination of the tentacular club, which is without hooks, prevents its affiliation to this species.

The fins of these two specimens are larger than in the figures that have been given of the species. This may be due to a simple sex difference.

A water-color sketch of the fresh animal by M. Tinayre shows that the fins are colorless, the head and arms are lilac and the viscera are slightly pinkish.

LARVAL FORMS

The following series of small cephalopods refer to larvae or juveniles frequently caught in bad condition. Their identification becomes doubtful because, in such stages, the generic and specific characters are very poorly developed.

I thank Mr. G. Pfeffer of the Hamburg Museum to whom I had sent most of these specimens and who was kind enough to give me his opinion on their identification.

Larvae possibly attributable to genus
Pterygioteuthis *or* Pyroteuthis

Expedition of 1904: Stn. 1715, depth from 0 to 1,000 m. Large Richard net. Close to Tenerife. A larvae approximately 9 mm long. Stn. 1856, depth from 0 to 3,259 m. Large Richard net. Azores. A larvae approximately 4 mm long.

Expedition of 1905: Stn. 2058, depth from 0 to 500 m. Large Richard net.

A larva approximately 8 mm long, with tentacles with a few large suckers as in the larva of *Octopodoteuthis* in the illustrations by Chun (**9**, Pl. XVII, Figs. 1 to 10). However, the rest of the body does not conform to this species. Pfeffer has presented something similar (**70**, Pl. XIX, Fig. 11, *Octopodoteuthis sicula* Rüppell). Rudimentary luminous organs are visible on the ventral part of the eyes but they are not yet pigmented. The suckers of the arms have an irregular, chitinous outline indicating the stage of transformation to hooks.

Stn. 2130, depth from 0 to 3,000 m. Large Richard net (Pl. XII, Fig. 8).

A juvenile approximately 7 mm long. I considered it necessary to photograph it to show the arrangement of the luminous organs on the anterior surface of the eyes.

Stn. 2168, depth from 0 to 2,000 m. Large Richard net. Juvenile—approximately 4 mm long.

54 I considered it necessary to present only one illustration of these various juveniles, since the others correspond to already described stages of development.

Genus *Enoploion* G. Pfeffer, 1912

Enoploion eustictum G. Pfeffer, 1912
(Pl. XIII, Figs. 5 and 6)

1912. *Enoploion eustictum* G. Pfeffer (**70**), p. 124 and 176

Expedition of 1909: Stn. 2902, depth from 0 to 1,800 m. Large Richard net.

This catch included six young cephalopods from 5 to 7 mm in length. I sent them to Mr. G. Pfeffer who was kind enough to examine them and who linked them to the species *Enoploion eustictum* that he had established in 1912. I would not have been sure enough of this identification to publish it, had not Mr. Pfeffer conveyed it to me himself; his genus *Enoploion* was in fact established only on the study of juveniles shown in Figs. 15 and 16 of his Plate XV. I would have considerred them as juveniles of *Abraliopsis*, but I have no hesitation in adopting the opinion of Mr. Pfeffer who has already studied this question and who is the only one to have seen representatives of this new genus. I would not be astonished, however, if these young cephalopods are later affiliated to a species already known in the adult stage. Their features, particularly those of the fins, have not yet become well-defined; the shape of this organ, the arms and suckers probably have not yet reached their definitive stage.

I have considered it necessary to give photographs of two of these young cephalopods. The others completely resemble the forms already published by Pfeffer. They will serve as a document for later reconstructing the series of juvenile stages of this species when more catches will complete their identification. One is of an individual 6.5 mm long and the other of an individual 5.5 mm long.

Genus *Octopodoteuthis* Rüppell, 1847

Octopodoteuthis sicula Rüppell, 1847

1847. *Octopodoteuthis sicula* Rüppell (**75**)
1845. *Verania sicula*, Krohn (**46**), p. 38

Expedition of 1905: Stn. 2172, surface. Fine meshed net. 9.50 p.m. to 10.30 p.m. Azores.

55 A very young embryo, 3.5 mm long, with head retracted into the mantle; only the tips of the eyes and terminals of the tentacles can be distinguished.

Family HISTIOTEUTHIDAE

Genus *Stigmatoteuthis* G. Pfeffer, 1900

Stigmatoteuthis chuni G. Pfeffer, 1912
(Pl. XIII, Fig. 3)

1912. *Stigmatoteuthis chuni* G. Pfeffer (**70**), p. 286
1910. ex *Calliteuthis hoylei* Chun (**9**)
? 1880. ex *Calliteuthis reversa* Verrill (**85**)

Expedition of 1903: Stn. 1583, depth 1,490 m. Trawl.

The creation of the genus *Stigmatoteuthis* is the result of the complete revision of the family Histioteuthidae by G. Pfeffer. The old species like *Histioteuthis bonnelliana* and *Histioteuthis rüppelli* to which small forms resembling them were affiliated, were no longer regarded by G. Pfeffer, according to Chun to be adequate to continue accommodating all these new forms. In 1910 (Valdivia Expedition) Chun had already begun placing these new forms elsewhere; but Pfeffer did not accept Chun's division and the nomenclature was once again changed.

This *Stigmatoteuthis chuni* is the result of these revisions.

When I examined this specimen in 1903 I considered it a young *Calliteuthis reversa* Verrill 1880. When Chun's memoir was published, I thought it would be more proper to assign it to *Calliteuthis hoylei* Chun,

1910. However, Pfeffer had created a new genus *Stigmatoteuthis* when revising the family Histioteuthidae in 1900. He restored the species *Histiopsis hoylei* Goodrich, 1896 and renamed it *Stigmatoteuthis hoylei* Pfeffer, 1900. To Pfeffer, the *Calliteuthis hoylei* of Chun was affiliated to his genus *Stigmatoteuthis.* Thus there were two *Stigmatogeuthis hoylei* for two different animals, Goodrich 1896 and the other Chun 1910. Pfeffer thus named *Calliteuthis hoylei* of Chun, which did not have priority, *Stigmatogeuthis chuni* Pfeffer. There are now in the genus *Stigmatoteuthis* of Pfeffer a *S. hoylei* Goodrich (ex *Histiopsis hoylei* Goodrich) and a *S. chuni* (ex *Calliteuthis hoylei* Chun). This is obviously an extremely complicated situation. Moreover, I am not at all convinced that we are in the presence of a final solution. Who knows how many more "turns of a waltz" the species of these genera will make!

56 Whatever the case may be, in order to conform with Pfeffer, who is certainly the most competent naturalist on these species, I assign my specimen to *Stigmatoteuthis chuni* G. Pfeffer.

However, this is without a firm conviction, and I ask myself whether *Histiopsis hoylei* Goodrich, *Stigmatoteuthis hoylei* of Pfeffer, *Calliteuthis hoylei* of Chun, *Stigmatoteuthis chuni* of Pfeffer and others, are not simply all synonyms, and whether or not all these may one day be reassigned to the old *Calliteuthis reversa* Verrill.

This small cephalopod has the following dimensions in millimeters.

Total length including tentacles	71
Length of mantle (measured dorsally)	21
Length of tentacle	45
Length of 1st arm (dorsal)	24
Length of 2nd	26
Length of 3rd	24
Length of 4th (ventral)	30
Width of the two fins	14

Luminous organs are abundant on the ventral surface, decrease on the sides of the mantle, but there are still a few on the dorsal surface, particularly two on each side of the dorsomedial line, 4 to 5 mm from the edge of the mantle.

I do not consider it necessary to give a new drawing of this specimen. Pl. XVIII, Fig. 1 of Chun gives an excellent ventral view.

Genus *Calliteuthis* Verrill, 1889

? *Calliteuthis* sp.
(Pl. IX, Fig. 10)

Expedition of 1905: Stn. 2001, depth from 0 to 1,500 m. Large Richard net. Corsica.

A juvenile, approximately 5 mm long, corresponds to Fig. 2, Pl. XX of G. Pfeffer (**70**). I have reservations on its assignment to the genus *Calliteuthis,* because the main characters are not adequately discernible. For the meantime, I keep this designation which reflects the opinion of G. Pfeffer. No traces of luminous organs are visible.

I considered it necessary to photograph this larva as a matter of record, but experienced great difficulties because alcohol preservation had turned the specimen almost black.

57 Family BENTHOTEUTHIDAE G. Pfeffer, 1912

Genus *Benthoteuthis* Verrill, 1885

Benthoteuthis megalops Verrill, 1885
(Pl. XIII, Fig. 4)

1885. *Benthoteuthis megalops* Verrill (**89**), p. 402
1885. *Bathyteuthis abyssicola* Hoyle (**25**), vol. I
1900. *Bathyteuthis abyssicola*, G. Pfeffer (**69**), p. 73

Expedition of 1909: Stn. 2876, depth from 0 to 1,000 m. Large Richard net; in the open waters off the Corogne.

The specimen captured at this station was a juvenile measuring 7.5 mm from the mouth to the posterior end of the body. The specimen is colorless; however, M. Tinayre observed a live form and the surface of its arms was bright red; the skin between the eyes was reddish-orange; the edge of the mantle was bright red and faded toward the posterior end of the body; above the fins a red circle surrounded the lower part of the mantle. The specimen is well preserved and I could photograph its ventral side (Pl. XIII, Fig. 4). In this young stage the arms form a regular crown. They are all of the same size and their end is turned down like a crozier toward the base as in the adult.

Benthoteuthis sp.
(Pl. XII, Fig. 9)

Expedition of 1908: Stn. 2695, depth from 0 to 2,595 m. Large Richard net. Balearic Islands.

One specimen, 3.5 mm long, resembling the form represented by Pfeffer (**70**, Figs. 14 and 15, Pl. 27). The tips of the tentacles are turned outward and form a sort of sole overlaid with four rows of small suckers. Since this specimen is slightly younger than the form shown by Pfeffer I have represented it (Pl. XII, Fig. 9) as a record before it is placed in a series to be reassigned.

58 Genus *Ctenopteryx* Appellöf, 1890

Ctenopteryx sp.
(Pl. XIII, Fig. 1)

Expedition of 1905: Stn. 2130, depth from 0 to 3,000 m. Large Richard net.

A very young specimen approximately 7 mm long, almost corresponding to Fig. 18, Pl. 27 cf Pfeffer. It is impossible to make out to what species it belongs, but it is easy to make out the genus from the fins which are filamentous as clearly seen on this embryo. I considered it necessary to photographically reproduce it as it is a very rare record. To date no other larval form is known except the one described by Pfeffer. The arms are all noticeably equal and covered with suckers. The tentacles are slightly longer than the arms and have a short but clearly marked club. The fins resemble two small combs and are perfectly characterized; they are arranged like two stems diverging from the tip of the body where they unite toward the mantle edge. The longest filaments of the fins are closer to the posterior end of the body; the shorter, scarcely visible, are nearer the head.

Family BRACHIOTEUTHIDAE G. Pfeffer, 1912

Genus *Tracheloteuthis* Steenstrup, 1881

Tracheloteuthis sp.
(Pl. XIII, Fig. 2)

Expedition of 1909: Stn. 2905, depth from 9 to 1,200 m. Large Richard net.

A young individual corresponding to the stage shown by G. Pfeffer (**70**) in Figs. 7 and 8 of his Pl. 26. The specimen is 11 m long from the mouth to the posterior end of the body. This animal probably resembles *Tracheloteuthis riisei* Steenstrup but I could not prove it due to its inadequate development. Pfeffer considers genus *Tracheloteuthis* of Steenstrup to be a subgenus of *Brachioteuthis* and he therefore mentions the name of this species as *Brachioteuthis* (*Tracheloteuthis*) *riisei* Steenstrup.

I have given the illustration of this young *Tracheloteuthis* to show the great difference in the development
59 of the tentacles and arms. The latter are very small in relation to the former; the dorsal and ventral arms are much smaller than the lateral arms. The head is dorsal, the eyes slightly protruding, the fins are small.

Family IV. OMMATOSTREPHIDAE Steenstrup, 1861

Genus *Ommatostrephes* d'Orbigny, 1835

Ommatostrephes sagittatus Lamarck, 1798

Expedition of 1909: Stn. 2929, surface. Caught in shrimp net under the searchlight. One individual 25 cm long, three individuals 8 to 10 cm long and a small individual 3 cm long.

Expedition of 1910: Stn. 3008, surface. A large specimen 1.02 m long attracted by the signal light of the ship and killed with a spear.

Genus *Illex* Steenstrup, 1880

Illex illecebrosus (Lesueur), 1821

Expedition of 1903: Stn. 1447, depth 130 m. Beam trawl. In the offshores of Cordouan. One individual 35 cm long. Stn. 1455, depth 358 m. Gulf of Gascony. Beam trawl. One individual 33 cm long, including the tentacles.

Genus *Todaropsis* Girard, 1890

Todaropsis eblanae (Ball), 1841
(Pl. VIII, Fig. 7)

1841. *Loligo eblanae* Ball (**3**), p. 363
1890. *Todaropsis veranyi* Girard (**19**), p. 204
1900. *Todaropsis eblanae*, G. Pfeffer (**69**), p. 179

Expedition of 1902: Stn. 1442, depth 152 m. Beam trawl.

One average size male, approximately 30 cm in length. The two ventral arms are hectocotylized, but

their transformation is very different from the common mode; I thought it would be useful to depict it (Pl. VIII, Fig. 7). It may be observed that the right arm is far more strongly modified than the left arm. On the left arm only the basal suckers are changed into triangular scales. From the seventh sucker to the tip of the
60 arm, the suckers are normal in appearance. On the right arm eight pointed scales can be seen from the base. They are hard and the first scale is very large; then a large lamella in the form of a fan is observed; beyond this all the suckers are transformed into foliaceous lamella and the edge of the arms in this region has a long flexuous membrane on the left side. This arrangement is extremely interesting and completely differs from the arrangement observed in other closely related species.

Expedition of 1903: Stn. 1455, depth 358 m. Gulf of Gascony. Beam trawl. One individual with a total length of approximately 35 cm. Stn. 1497, depth 208 m. Gulf of Gascony. Beam trawl. Two individuals, 23 and 26 cm long.

Expedition of 1905: Stn. 1994, depth 141 m. Near Monaco. Large beam trawl. A non-adult female with a total length of approximately 11 cm.

Genus *Stenoteuthis* Verrill, 1880

Stenoteuthis bartrami (Lesueur), 1821

1821. *Loligo bartrami* Lesueur (**52**), p. 90
1835. *Ommastrephes bartrami*, d'Orbigny (**63**), p. 55
1880. *Stenoteuthis pteropus* Verrill (**85**), p. 228
1900. *Stenoteuthis bartrami*, G. Pfeffer (**69**), p. 180

Expedition of 1904: Stn. 1746. Canaries. One juvenile approximately 10 cm long found on the bridge.

Expedition of 1905: Stn. 2184, surface. San Miguel, Azores.

One specimen approximately 21 cm long, caught under an electric searchlight.

Genus *Rhynchoteuthis* Chun, 1903

Several of these small cephalopods, probably in an embryonic stage of some Ommastrephidae, were captured at various stations.

Expedition of 1905: Stn. 2073, surface. Fine-meshed net. Stn. 2078, surface. Fine-meshed net.

Expedition of 1908: Stn. 2688, depth from 0 to 2,025 m. Large Richard net.

Expedition of 1909: Stn. 2905, depth from 0 to 1,200 m. Large Richard net.

Expedition of 1910: Stn. 3053, surface. Fine-meshed net.

These embryos resemble the descriptions given by various authors, and it is not necessary to recapitulate them here.

61 Family CHIROTEUTHIDAE Gray, 1849

Subfamily Chiroteuthinae G. Pfeffer, 1912

Genus *Chiroteuthis* d'Orbigny, 1839

Subgenus *Doratopsis* de Rochebrune, 1884

Subgenus *Planctoteuthis* Pfeffer, 1912

Chiroteuthis (*Doratopsis* or *Planctoteuthis*) (sp. ?)

Expedition of 1905: Stn. 2022, depth from 0 to 4,000 m. Large Richard net.

Two very young, small specimens, in very bad condition, which, in the opinion of Mr. Pfeffer, could not be identified with great certainty. They were 5 to 8 mm in length; they could not be accurately measured because they were considerably mutilated.

Genus *Doratopsis* de Rochebrune, 1884

Doratopsis sp.

Expedition of 1910: Stn. 3039, depth from 0 to 3,660 m. Bouree net rapidly towed.

This specimen is only a head without eyes of an average-sized individual that I place in this genus with great reservations.

Subfamily Mastigoteuthinae G. Pfeffer, 1912

Genus *Mastigoteuthis* Verrill, 1881

Mastigoteuthis magna L. Joubin
(Pl. IX, Figs. 6 and 7; Pl. X; Pl. XI, Figs. 1 to 5)

1913. *Mastigoteuthis magna* L. Joubin (**45**), p. 1 to 11

Expedition of 1905: Stn. 2108, depth 3,465 m. Sargasso Sea. Triangular bow net.

62 This very beautiful cephalopod was captured in a triangular bow net lowered to 3,465 m. When the bow net was opened, a separated tentacle was found near it. The severed surface corresponded well with the stump still adhering to the tentacular crown of the animal. The other tentacle was absent. Moreover, the color and nature of the skin of the tentacle corresponded well with that of the cephalopod. There is, therefore, no doubt that the tentacle belonged to the specimen.

From this, I initially thought I had found the solution to a small problem which for many years had excited the curiosity of naturalists following the oceanic cruises of HRH, the Prince of Monaco, viz., to which species of cephalopods would the tentacles belong that are frequently found coiled around the sounding lines. No one has succeeded in catching the animal.

In 1895 I described this "unknown tentacle" and particularly its club with small suckers at its tip. Seeing, this time, a tentacle totally resembling the cephalopod to which it most probably belonged, I thought I had finally captured the cephalopod to which I could affiliate the numerous tentacles gathered over 20 years. However, on closer examination, I realized that the tentacle caught with the cephalopod in the bow net was not the same as the one frequently encountered on the sounding lines.

As will be seen in the following description, the tentacle of *Mastigoteuthis magna* has a club almost entirely covered with suckers but they are so small that they are invisible to the naked eye. On the contrary, the indeterminate tentacles have a shorter, broader club with numerous suckers visible to the naked eye. Moreover, the horny ring of the suckers on the indeterminate tentacles is dentate, whereas the suckers of *Mastigoteuthis magna* are completely smooth. For these reasons, the identification of the indeterminate tentacles remains unresolved. However, since my publication in 1895, several species of *Mastigoteuthis* with similar but not identical tentacles have been described, thus providing better grounds for stating that the indeterminate tentacles belong to cephalopods of the genus *Mastigoteuthis.*

Another acceptable hypothesis is that the difference between the tentacular clubs of the two species is merely due to a difference in sexes. It is impossible to confirm this, since I have had only a single specimen of *Mastigoteuthis magna* that I did not dissect because I wanted to keep it intact. Therefore, I do not know its sex. More material will have to be obtained before we can know whether this suggestion is correct. Until then, the tentacle of *Mastigoteuthis magna* and the tentacles of the *Mastigoteuthis* of the unknown species must remain as two distinct species.

All new records have been referred to in this brief preliminary discussion and now the specimen will be described. After studying the bibliography it appeared almost certain that the animal belongs to the genus
63 *Mastigoteuthis* but not one of the species described to date conforms completely to the animal under discussion.

First, all species with luminous organs described so far must be eliminated because this specimen has none. Among the others, *Mastigoteuthis cordiformis* Chun is also eliminated since the shape of the fins is very different and the suckers have a dentate horny circle which is smooth in my specimen. For the same reason, the specimen cannot be affiliated to *M. dentata* Hoyle which is furthermore inadequately described. It also differs from *M. levimana* Lönnberg which has a dentate horny ring around the suckers and ventral arms with eight suckers in a single line commencing from the mouth before the suckers become arranged in two rows. In my specimen the arrangement of two rows of suckers begins right at the mouth.

For these various reasons, I consider this to be a new species and I name it *Mastigoteuthis magna* since, for the present at least, it is the largest known form of *Mastigoteuthis*.

M. Tinayre made a water-color sketch of the animal immediately after its capture, and based on it, Miss Vesque has made the beautiful color drawing in Pl. X.

The entire animal must have been brick red mixed with violet. Unfortunately, the skin has suffered greatly either during handling or from damage caused while in the bow net, because a large part of its epidermis has disappeared; moreover, it had coiled one of its tentacles tightly around its head leaving a white mark where the skin pulled by it was damaged.

The main measurements of the body of this cephalopod are given below (in mm):

Total length from the tip of the ventral arm	382
Length of the first arm (dorsal)	80
Length of second arm	105
Length of third arm	82
Length of fourth arm (ventral)	172
Length of the mantle (ventral)	151
Length of the mantle (dorsal)	160
Length of the fin	105
Width of the fins	127
Width of head, at eye level	34
Width of the head, below the eyes	43
Length of the head	42

64 The formula for the arms according to these measurements is thus 4, 2, 3, 1. The characteristics of the different external organs can be summed up as follows.

The mantle is in the shape of a gradually tapering cone to the posterior end of the body, broken at the level of the fins; one end of the gladius projects like the brown rib of a feather. It is thus impossible to know the length of the posterior extremity of the mantle beyond the fins. However, this probably could not have been very long. The color of the skin of the mantle is dark brownish-violet, becoming lighter toward the lower tip. The skin is smooth and is scattered with a large number of very small chromatophores.

Dorsally, there is a distance of 54 mm between the mantle margin and the anterior insertion of the fins which measure 116 mm long along the dorsal length of the mantle; thus the ratio 54/116 is approximately one-third part of the mantle not united to the fins as against two-thirds for the united part. Due to the mutilation of the posterior end of the mantle a precise figure could not be given.

The mantle opening is approximately 42 mm in width; its upturned lateral and ventral edges do not permit accurate measurement.

The mantle-cephalic adhesive cartilages are very distinct. The lateral two are in the form of a vertical rectilinear ridge on the mantle, narrow anteriorly and slightly wider posteriorly. These fit into the two auricular-like grooves with a reddish-violet outline and their concavity exactly agrees with the relief of the mantle ridges.

The unpaired dorsal mantle ridge and the cephalic groove are parallel to each other. On the head the two vertical protruding crests are separated anteriorly by a narrow furrow and bordered by two broad furrows. The entire structure is surrounded by a thin membrane with a violet-colored border. On the mantle, the corresponding ridge, also surrounded by a membrane with a violet-colored border, comprises a thin median protuberance bordered by two strong grooves entering the cephalic furrow. The ventral grooves are 11 mm long for a maximum width of 7 mm; the dorsal groove is 17 mm long with a width of 7 mm.

The funnel has a large collar extending around the head to the dorsal nuchal cartilage. Its median tube is short and curved and does not reach the level of the eyes; its length is only 18 mm.

The lateral areas are well developed as are the pillars that carry the funnel locking apparatus. The valves obliquely descend along the length of the edge of the dorsal adhesive cartilage and are inserted into a triangular, reddish-brown membrane.

The fin is strong, very thick, and muscular; it is 105 mm long with a maximum width of 127 mm. It is thus slightly elliptical. Its tip probably slightly extends below on the point of the gladius which extends beyond the mantle. It is entirely covered by a fine reddish-brown violet skin, unfortunately in shreds, slightly lighter colored than the ventral sac.

65 The head has a rather peculiar shape due to its length and narrowness at the level of the insertion of the arms. It basally bulges below the eyes on each side, above each of the two valves of the funnel. These two large cheeks are pouches in which the two eyes are lodged. At the time of capture I initially thought, as is frequently the case, that the eyes were absent and that the two orbits were empty. On closer examination I found the two eyes quite intact inside these large pouches. They are very large, flat, and the opening of the orbital cavity is rounded without a well-marked sinus. The inside of the orbital cavity is violet tinted.

Just above the orbit, emerging from the cutaneous fold uniting arms three and four, a small fragment of tentacle can be seen still adhering to the head. This stump, which could be about 5 or 6 mm long, has a sharply severed surface. Its skin is brownish-violet and its shape is cylindrical; it is muscular and in size, shape and color seems to fit the cut-surface of the isolated tentacle found with it in the bow net.

The tentacle found in the bow net with this cephalopod is 43 cm in length; its overall color is light brownish-red. It is cylindrical and gradually terminates in a fine tip. Its width does not exceed 4 mm, and that is its most swollen portion; everywhere else it is approximately 3.5 mm and becomes filiform at the tip. Except for the 4 cm portion of the tentacle between the head and club, all the rest, i.e., 39 cm, constitutes the club. It is so little distinct from the rest of the tentacle, with no membrane surrounding the cupuliferous region so that at first glance it seems that the tentacle has no suckers. Close examination reveals a slightly granular region on the surface of the skin and with microscopic examination it is found that the granular texture is due to thousands of extremely small suckers. The surface covered by these suckers starts very close to the head like a narrow band which gradually broadens near the tip of the tentacle. In the middle of this club, 3/4 of the surface of the tentacle is covered with suckers and toward its tip, the entire surface is hidden under this carpet of small yellow dots.

Chun has shown something similar for his *Mastigoteuthis cordiformis.* Fig. 8 of his Pl. XXXV depicts a brachial crown with one of the tentacles covered with small suckers starting near the middle of the club and terminating in a point, as in our species. The proportion of the club to the tentacle is not the same, since almost the entire tentacle of my species is transformed into a club; whereas, only half is transformed in the Chun species. On the other hand, the suckers of *M. cordiformis,* although all small, are much larger and,

66 even though numerous, are not innumerable as in *M. magna*. In the same Pl. XXXV, Fig. 16, Chun has shown a part of the club of *M. glaukopis* which gives a better idea of the profusion of suckers in *M. magna*. It is difficult to represent the suckers of this tentacle in their actual dimensions, since they are not visible to the naked eye and the tentacle resembles a smooth cord. A very small membrane surrounds the region of the club, but it is so faintly marked that it seems to be almost non-existent. This was not visible to the naked eye because, although the specimen was preserved as soon as it was captured, it was in rather poor condition and part of its epidermis had disappeared.

These minute suckers are probably the smallest that have ever been detected in any adult cephalopod. Even those of *Mastigoteuthis glaukopis* represented by Chun in Fig. 16 of his Pl. XXXV with 20 × magnification, appear to be very large in comparison to the suckers of *M. magna*.

The suckers have a rough, transparent peduncle fixed on the surface of the tentacle and supporting a chitinous apparatus resembling a small lantern. The horny part is a lovely yellow color and consists of two well-defined parts: one in the shape of a vaguely cubic case into which the soft tube penetrates through one of its faces; the other is a broad funnel with a narrow tube inserted into one of the faces of the case at right angles to the base. This funnel is absolutely round, broad and shallow, and is marked with polygonal ornaments arranged like a mosaic around the tubular orifice. The first row of this mosaic consists of some dozen pentagons arranged around the central hole; in the second row, the pentagons are smaller and more numerous. Finally, at the edge of the funnel a narrow row of very small polygons surrounds this entire rosette. A thin smooth horny line encircles the whole like a cut scale. Each of the polygons contains a protuberance like a pearl or tubercle in its center. There is no protruding denticle at the edge or center of the funnel and the tubercles of the polygons are very slightly elevated. These peculiar suckers differ greatly from those of all other species of *Mastigoteuthis*, as well as those of the unknown tentacle I described in 1900.

No papilla (olfactory?) is present on the skin below the eyes as is found in various other species of the same genus.

The arms are very unequal as is evident from table of measurements. The ventral arms are much longer than the others which is the rule in all species of the genus *Mastigoteuthis*.

The skin of these arms is lighter on the outer surface than on the inner surface which supports the suckers where it is dark violet. The skin is soft, as though impregnated with water. These arms are flaccid and only those of the ventral pair have a well-marked muscularity.

There is no projecting membranous ridge on the skin, at least as far as can be detected from the poor condition of the skin. The surface with the suckers is marked by a slight elevation in the skin but this cannot
67 be qualified as a membranous ridge even though it is slightly raised. The section of the arm shows that it is not completely rounded; the surface with the suckers is slightly flattened, but the opposite surface is slightly swollen, almost keeled on the ventral arm.

The suckers are arranged in two, rather widely spaced rows, on the first three arms. There are two rows on only half the length of the ventral arms; they are in a slightly wavy single row on their tips.

The ventral arm has 18 pairs of large suckers in two separate rows. From the 19th sucker to about the 80th they are in a single line. Thereafter there are approximately 12 very small suckers, but I could not be definite about the number as the tip of this arm was damaged.

The third arm has 56 pairs of suckers in two distinct rows; the end suckers are very small. The second arm has 60 distinct pairs of suckers and, moreover, some very small ones at the tip; I could not count the exact number. The first arm has 65 distinct pairs of suckers and another five or six very small ones at the tip.

The brachial suckers are almost spherical; they progressively increase in size from the base of the arms toward the 12th or 13th pair, thereafter they maintain the same diameter for about 15 rows after which they diminish in size to the tip of the arm. Their peduncle is short; at first glance they seem to be almost sessile; their surface is pigmented violet-red, slightly lighter than the skin of the arm; the peduncle is colorless. Their horny ring is dark brown with numerous lighter colored bands; it is completely without denticles, a feature that is not found in any other species of *Mastigoteuthis*; there is not even the suggestion of small undulations on the edge of the absolutely smooth opening. The inside of the sucker is white. The opening is almost round but slightly longer than broad.

The arms are connected to the peribuccal membrane by bridles issuing from the dorsal margin of the base of the first two arms and from the ventral margin of the base of the last two arms. The two dorsal bridles join on the dorsal median line at the peribuccal membrane; the same holds for the two ventral bridles. The contour of this membrane represents a hexagon in which the angles are formed by the insertion of brachial bridles. At each angle a protrusion of the edge of the membrane forms a sort of triangular crest, the remainder of the edge of the membranous edge of the lip is slightly denticulate; its entire surface is smooth; it is dark violet like the skin of the arms.

Inside this peribuccal membrane the lip can be seen as a large white papillose ring and radial folds going deep toward the tip of the horny beaks.

Nothing much can be said about the gladius, since I did not wish to completely spoil the sole specimen of this already damaged cephalopod which was to be deposited in the collection of the Museum of Monaco.

68 *Mastigoteuthis* sp.

This refers to the fragments of the "unknown tentacle" discussed under *Mastigoteuthis magna* Joubin. This list is compiled from stations where these fragments were observed.

Expedition of 1902: Stn. 1307, surface. Azores. Stn. 1325. On baited line. Azores. Stn. 1408. On the sounding lines. Azores.

Expedition of 1904: Stn. 1751b. On the sounding lines. Stn. 1893. On the sounding lines. Parts of the Gorringe Bank.

Expedition of 1905. Stn. 2123. On the bow net cable at Stn. 2108. Sargasso Sea. Stn. 2148. On the sounding lines. Sargasso Sea.

Expedition of 1906: Stn. 2350b. On the sounding lines lowered to 2,600 m over a sea bed of 4,800 m. Coast of Portugal.

Mastigoteuthis sp.

Expedition of 1902: Stn. 1334, depth 1,900 m. Azores.

One specimen in very poor condition which seemed to belong to either *Mastigoteuthis flammea* Chun or *M. cordiformis* Chun. However, it is not possible to identify it with certainty.

Expedition of 1905: Stn. 2130, depth from 0 to 3,000 m. Large Richard net.

This again refers to a small specimen in very poor condition which seems to be affiliated either to this genus or to *Chiroteuthis*.

Family CRANCHIIDAE G. Pfeffer, 1912

A. Subfamily Cranchiinae G. Pfeffer, 1912

Genus *Leachia* Lesueur, 1821

Leachia cyclura Lesueur, 1821

1821. *Leachia cyclura* Lesueur (**52**), p. 90, Pl. VI

Expedition of 1901: Stn. 1235, surface. Waters off Madeira.

Expedition of 1902: Stn. 1409, surface. Azores. Four specimens.

Expedition of 1904: Stn. 1799, surface. Sargasso Sea. Two specimens. Stn. 1850, surface. Several specimens. Stn. 1880, surface. Azores. Several specimens.

69 Expedition of 1905: Stn. 2178, surface. Azores. One specimen. Stn. 2186, surface. Found in the hold of *Hirondelle*. Two specimens.

Genus *Cranchia* Leach, 1817

Cranchia scabra Leach, 1817

1817. *Cranchia scabra* Leach (**49**), p. 40

Expedition of 1904: Stn. 1797, depth from 0 to 2,000 m. Canary Islands. Large Richard net.

One deformed specimen, approximately 6 mm in length. Another the same size, in poor condition, which leaves doubt on its identification.

Expedition of 1905: Stn. 2153, depth from 0 to 2,000 m. Canary Islands. Large Richard net.

One young specimen with a total length of 8 mm, including the arms. It is in good condition and has the appearance of a small semitransparent balloon after preservation in formalin.

Subfamily Taoniinae G. Pfeffer, 1912

Genus *Taonius* Steenstrup, 1861

Taonius pavo (Lesueur), 1821

1821. *Loligo pavo* Lesueur (**52**), p. 96
1839. *Loligopsis pavo*, Férussac et d'Orbigny (**63**), p. 321
1861. *Taonius pavo*, Steenstrup (**77**), p. 83
1881. *Desmoteuthis hyperborea* Verrill (**86**), p. 302

Expedition of 1902: Stn. 1308, surface, Azores.

A very good specimen was captured at the surface. Its total length is approximately 40 cm. Since it differs in no way from the specimen I described and illustrated in 1900, it is pointless to give a redescription. Its color, size and ratios of the parts, are identical.

70

Genus *Galiteuthis* Joubin, 1898

Galiteuthis armata Joubin, 1898

(Pl. XIV, Fig. 1)

1898. *Galiteuthis armata* Joubin (**35**), p. 279
1910. *Galiteuthis* (*Taonidium*) *suhmi* Chun (**9**), p. 382
1912. *Galiteuthis armata*, Pfeffer (**70**), p. 731

Expedition of 1910: Stn. 3045, depth from 0 to 1,400 m. Bouree net rapidly towed. Between Cape St. Vincent and Gibraltar. Two specimens.

Two young specimens in poor condition. The main measurements of the better preserved cephalopod are given here. The measurements of some parts could not be given since the tentacular clubs and eyes were mutilated or completely severed.

The specimen has the general typical characters of whiteness, transparency, arrangement of fins, suckers, and funnel. There is nothing to add to the earlier descriptions.

The total length of the specimen is 113 mm including the tentacles and 91 mm without them. The tail from where the fin arises is 39 mm. The only entire tentacle is 25.5 mm long.

The second individual is slightly smaller, and since the head is separated from the body, the eyes absent and the arms are tightly entwined it has not been possible to take any measurements.

Expedition of 1905: Stn. 1991, surface. On arriving at Monaco. Fine-mesh net (Pl. XIV, Fig. 1).

One very young larva, exactly corresponding to Figs. 9 and 10 of Pl. LXI of Chun (**9**, Expedition of the Valdivia).

The body, measured from mouth to fin, is a little less than 7 mm; the tentacle is approximately 5 mm; the arms are very rudimentary and resemble small peribuccal papillae. The eyes are very prominent and larger than the buccal bulb.

Stn. 2118, depth from 0 to 3,465 m. Beam trawl, Sargasso Sea.

One young *Galiteuthis armata*, approximately 10 cm in length from the tip of the tail to the mouth. A watercolor sketch by M. Tinayre at the time of its capture shows a peculiar red dot below each eye and a black ring with a lilac-colored ring on top of it at the boundary between the ocular capsule and the lens.

This specimen is probably the same as the *Galiteuthis phyllura* described by S. Berry in 1911 in his *Review of the Cephalopods of Western North America.*

71 **Genus *Teuthowenia* Chun, 1910**

Subgenus *Teuthowenia* Chun, s. str.

Teuthowenia megalops (Prosch), 1847

1847. *Owenia megalops* Prosch (**71**), p. 47
1850. *Cranchia megalops*, Mörch (**59**), p. 57
1900. *Owenia megalops*, G. Pfeffer (**69**), p. 193
1912. *Teuthowenia megalops*, G. Pfeffer (**70**), p. 742

Expedition of 1904: Stn. 1715, depth from 0 to 1,000 m. Ilheus Selvagens. Large Richard net. One specimen with 13 mm long body, excluding the head; in rather poor condition. Stn. 1749, depth from 0 to 2,500 m. Between Madeira and the Canary Islands. Large Richard net. One young specimen in poor condition, which makes identification doubtful. The body is approximately 7 mm long. Stn. 1849, depth from 0 to 3,000 m. Azores. Large Richard net. One young individual in good condition, approximately 8.5 mm long. The chromatophores form two or three circles on the mantle as shown in Fig. 17 of Pfeffer. Stn. 1874, depth from 0 to 2,000 m. Large Richard net. One specimen 19 mm long, excluding the arms. The tentacles are in poor condition. The body of the fresh animal is semitransparent with numerous chromatophores resembling small reddish colored dots on the ventral side, particularly on top.

Expedition of 1905: Stn. 2187, depth from 0 to 2,500 m. Azores. Large Richard net. One specimen in good condition whose body is 1 cm long and the tentacles 1 cm long.

Expedition of 1910: Stn. 3039, depth from 0 to 3,660 m. Bouree net towed rapidly. Waters off the Gorringe Bank.

In some details of the fin, this cephalopod resembles Pfeffer's form (**70**) (Pl. XLVIII, Figs. 17 and 18). In fact, its fin resembles that of *Helicocranchia pfefferi* Massy, but its other characteristics differ. Unfortunately, the tentacles are in poor condition. The specimen is 24 mm long from the mouth to the insertion of the fin; the tentacles are 42 mm long. This identification is the result of correspondence with Prof. G. Pfeffer.

72 **Genus *Hensenioteuthis* G. Pfeffer, 1900**

Hensenioteuthis joubini G. Pfeffer, 1900
(Pl. XIV, Figs. 2 and 3)

1900. *Hensenioteuthis joubini* G. Pfeffer (**69**), p. 143
1912. *Hensenioteuthis joubini*, G. Pfeffer (**70**), p. 747

Expedition of 1904: Stn. 1715, depth from 0 to 1,000 m. Canary Islands. Large Richard net.

One specimen approximately 1 cm in length from the mouth to the posterior end of the body. The tentacles are in rather poor condition. It strongly resembles the form illustrated by Pfeffer in Pl. XLVIII, Figs. 12 to 14 in his work on the cephalopods of the Plankton Expedition. It differs only in the absence of most of the chromatophores on the skin, probably due to defective preservation. However, one large one has persisted on the dorsal side of the ocular tubercle of each eye and a few on the ventral side of the funnel and mantle. The funnel is very large, protruding, and extends beyond the level of the mouth. The strong transverse musculature of the mantle can be distinctly seen. A tubercle is observed below each eye; it is unfortunately in such bad condition that it is difficult to define, but it is perhaps a luminous organ. The eye is not circular but slightly elliptical due to the presence of this tubercle.

Genus *Bathothauma* Chun, 1906

Bathothauma lyromma Chun, 1906
(Pl. XIV, Fig. 4)

1906. *Bathothauma lyromma* Chun (**7**), p. 86
1910. *Bathothauma lyromma*, Chun (**9**), p. 389

Expedition of 1910: Stn. 3039, depth from 0 to 3,660 m. Bouree net towed rapidly.

One specimen of a very young, almost embryonic form, unfortunately in very poor condition, but the features of the species are recognizable; length approximately 10 mm from the mouth to the posterior end. The specimen scarcely differs from the figures given by Chun for the adult. Only the head and part of the arms can be clearly seen; the shape of the eyes is absolutely characteristic. I have shown this specimen from highly enlarged photographs (Pl. XIV, Fig. 4) for comparison with Chun's figure.

73 **New genera of the family Cranchiadae**

The young cephalopods described here are affiliated to the family Cranchiadae, but since the specimens do not have their distinguishing features, and since I only have a single specimen of each that I cannot dissect, their diagnosis is incomplete. In any case this is merely a tentative description destined to undergo revision and completion when the occasion arises.

On the other hand, these young cephalopods do not conform to the rather rare juveniles of the family Cranchiadae described to date. I would prefer to place these forms in the anonymous group, but it is perhaps more advisable to follow the rules of nomenclature by attributing a name to them that expresses my doubts on the position that they will later occupy in classification.

Genus *Fusocranchia* nov.

Fusocranchia alpha n. sp.
(Pl. XIV, Figs. 5 to 9)

Expedition of 1904: Specimen A. Stn. 1800, depth from 0 to 1,000 m. Large Richard net. Halfway between the Azores and the Canaries (Pl. XIV, Figs. 5 to 7).

As already mentioned above, this cephalopod has been provisionally named after the Greek alphabet to express my doubts regarding its affiliation, since I do not have adequate records to make a detailed and comprehensive study.

This small animal, 9.7 mm in length from the beginning of the dorsal arms to the posterior end, is semitransparent, cigar-shaped and has a small head half buried in the large end. Only the eyes project from the mantle opening. From a bundle of very small arms two tentacles, 3 mm long, emerge from the center of the head.

A watercolor sketch by Mr. L. Tinayre at the time of its capture shows that the body is transparent, the head bright yellow and the eyes blue. Some yellow chromatophores are scattered over the dorsal side of the mantle. One large red chromatophore is located on each side of the head, below and inside the eyes.

Traces of these two large chromatophores are visible in the preserved animal.

The width of the semitransparent body is 3.5 mm at its widest level. A small dark spot, pink in the fresh animal, is visible on the body and corresponds to the visceral mass.

74 The margin of the mantle, with a 2 mm diameter of opening, has a depression in the middle of the dorsum and two depressions on the ventral side corresponding to the two positions of the funnel pouch valves. A darker gray line issues from each of these two ventral depressions and proceeds obliquely toward the posterior end of the body. It is slightly creased like a furrow; I saw no sign of papilliform structures. These two lines disappear toward the first 1/3 of the mantle.

The body posteriorly terminates in the form of a small papilliform tubercle which is traversed from front to back by a sort of membranous bridle. The fins are inserted in a ring on each side of this bridle; each is rounded and palette-shaped with a short, narrow sleeve. The sleeve is dorsally and obliquely inserted into the ring. The muscle fibers spread out fan-like. These fins are very small in relation to the size of the mantle. The enlarged part of the gladius is located between these fins. The gladius extends along the length of the dorsomedial line.

The arms are very slightly developed; the first dorsal, the third lateral and fourth ventral are reduced to small tubercles, very difficult to distinguish, which did not seem to have suckers. The second pair of arms was a little more developed and three pairs of small suckers and one terminal sucker could be distinguished. The tentacles are immeasurably longer than the arms—slightly more than 3 mm—with two rows of suckers from their base to the club which has four rows. The club is approximately one-third of the total length of the tentacle.

The suckers of the tentacles are very small and transparent. Those at the base have a slightly longer peduncle than those on the club. These are almost sessile, transparent, and have small islets of granulation. Each sucker has a horny, rather flat ring without denticles which communicates with the spherical cavity of this organ through a round, narrow neck. The ring walls have a slightly protruding ornamentation in the form of converging ribs which turn toward the orifice of the neck. A small line proceeding toward the edge of the ring runs between two consecutive ribs. These suckers are so small that their study is extremely difficult.

The funnel is wide and short; it extends above the level of the eyes; it is covered on top by its posterior wall recurved into a dome so that its opening is in the shape of a non-terminal ventral slit. A membranous bridle extends from one eye to the other above the funnel, under the tentacles and the rudimentary fourth arms.

No luminous organs were detected; I could not see below the eyes which are deeply buried in the mantle and therefore, I ignored that it existed. The large, circular eyes without stalks are prominent but obliquely directed upward and outward. They have a black spot visible through the skin.

75 I have not found in any work on cephalopods any form corresponding to this in the family Cranchiadae.

Specimen B. Stn. 1856, depth from 0 to 3,250 m. Large Richard net. Azores. (Pl. XIV, Fig. 8).

This is a very young specimen, 5 mm long. The features are so slight at this stage that precise identification is impossible. Nevertheless, I believe that this small animal can be assigned to the preceding species. The proportions of the arms, eyes, and funnel are similar. The black spot of each eye is distinctly visible. The body was deformed and has shrunk during preservation, but it can be identified from the shape of the mantle opening and the fin arrangement. The undeveloped tentacles are much shorter but the club has a similar disposition. The better preserved fin is notched like the heart in a deck of cards, its edge thin, free and transparent.

Specimen C. Stn. 1872, surface. Fine-mesh net (Pl. XIV, Fig. 9). The specimen described here is a very young larva with a total length of approximately 4 mm; only the tentacles are developed but are still very short. The other arms are scarcely distinct. The fins attached to the mantle are not visible. The mantle is large, ovoid. The eyes have the same shape and similar black spots as the two preceding specimens. With great reservations I associate this young larva to the same species as the two preceding forms.

Cranchiadae

Expedition of 1904: Stn. 1768, depth from 0 to 3,000 m. Canary Islands. Large Richard net.

One embryo 5.5 mm long, white, contracted, belonging to the Cranchiadae family, but it is impossible to identify it to the generic level. The specimen is too small, in poor condition and contracted by the preserving fluid. I did not photograph the animal.

Oegopsidae
(Pl. IX, Figs. 8 and 9)

Expedition of 1901: Stn. 1228, surface. Monaco Trench (Azores). Unknown embryos.

It is impossible to give a more precise identification of these young embryos found floating on the surface of the sea in a viscous egg-mass. Their body is short and rounded; the head is buried inside the mantle which
76 is broadly notched on the dorsal face. The fins resemble those of Sepiolidae. There are four rows of suckers on the lower half of the tentacles which are much more developed than the other arms. The arms of the fifth and third pairs surrounding the tentacles are at the stage of tiny stumps whereas the dorsal arms of the first and second pairs are well developed and have a few suckers in two rows. The eyes are completely buried in the mantle cavity; they are large and rounded but have a well-developed basal hemispherical mass. In view of the pelagic viscous egg-mass around these animals, the family must be separate from the Sepiolidae. They do not have the arm arrangement of the Histioteuthidae.

77 Unidentifiable Fragments

Expedition of 1902: Stn. 1440. Gulf of Gascony. Young unidentifiable oegopsid found in the stomach of a tuna.

Expedition of 1903: Stn. 1507. Unidentifiable gladii and one beak found in the stomach of an *Aphanopus carbo* caught on a trawl line lowered to 4,330 m.

Expedition of 1904: Stn. 1875, surface. Two beaks found in the stomach of a turtle. Stn. 1886, surface. Fragment of an arm with three luminous organs probably belonging to a species of family Histioteuthidae.

Expedition of 1908: Stn. 2689, surface. Eight beaks and six lenses of average size found in the stomach of a pilot whale, *Globicephala.*

Expedition of 1910: Stn. 3011. One lens found in the intestine of a *Centroscymnus coelolepis* caught on a bottom line below 1,680 m. Stn. 3050, surface. Six beaks (two large, four small) found in the stomach of a pilot whale, *Globicephala.*

Contents of Stomachs

Expedition of 1902: Stn. 1253, surface. Between Monaco and Corsica. In the stomach of a *Globicephala.*

Cephalopods were so digested and disintegrated that it was not possible to identify them. The following parts were found:

3 bodies of *Ommatostrephes* (?), 33 to 35 cm long.
2 similar bodies, approximately 30 cm long.
1 cone-shaped body, 20 cm long.
4 large brachial crowns.
3 smaller brachial crowns.
2 large heads.
Other remnants.

Expedition of 1908: Stn. 2689, surface. Stomach contents of a *Globicephala.* Body, without head or fins, belonging to a species of Ommatostrephidae, approximately 35 cm long. Imprints of the teeth of the cetacean could be observed on the body.

78 One brachial crown and a fin probably of the same cephalopod.

One body, a small brachial crown and two average size crowns so digested that they are absolutely unidentifiable.

Expedition of 1910: Stn. 3047. In the stomach of a dogfish shark, *Centrophorus calceus,* caught on a trawl line at 1,401 m.

2 large beaks.
2 small beaks.
1 large and 1 small lens.
Remains of a gladius.

Stn. 3050, surface. Contents of the stomach of a *Globicephala.*
3 heads of a large histioteuthid with arms.
3 brachial crowns of an unidentifiable decapod of average size.
1 brachial crown of an unidentifiable onychoteuthid.

1 body of an average size histioteuthid.
2 bodies of an average-size ommatostrephid.
1 unidentifiable fin.
2 small histioteuthids (?) in poor condition.
5 small brachial crowns of unidentifiable decapods.
29 horny beaks.
7 entire specimens of *Ommatostrephes* (?) approximately 40 cm long.
2 bodies of average-size *Ommatostrephes* (?).
1 head of an average-size *Ommatostrephes* (?).
1 brachial crown of an unidentifiable average-sized cephalopod.
Remains of approximately 12 gladii.

Tables of the Species Caught at Different Stations

EXPEDITION OF 1898

Station No.	Date	Locality		Depth in meters	Nature of bottom	Method of capture	Species caught
		Latitude	Longitude				
922	6 July	58°16′ N	5°48′15″ E	343	Mud	Trawl	*Octopus arcticus*
960	29 July	72°37′ N	20°00′15″ E	394	Mud and gravel	Trawl	*Octopus arcticus*

EXPEDITION OF 1899

Station No.	Date	Locality		Depth in meters	Nature of bottom	Method of capture	Species caught
		Latitude	Longitude				
1052	10 July	65°41′ N	9°30′15” E	440	Mud	Trawl	*Octopus arcticus* *Rossia glaucopis*

EXPEDITION OF 1900

Station No.	Date	Locality		Depth in meters	Nature of bottom	Method of capture	Species caught
		Latitude	Longitude				
1086	12 Aug.	51°53′ N	7°29′15″ W	64	Mud	Beam trawl	*Eledone cirrhosa*
1088	18–19 Aug.	Loch Juadh (West coast	[*sic*] of Scotland)	20		Trammel net	*Eledone cirrhosa*

EXPEDITION OF 1901

Station No.	Date	Locality		Depth in meters	Nature of bottom	Method of capture	Species caught
		Latitude	Longitude				
1118	12 July	29°06′30″ N	13°02′45″ W	1,098	Sandy mud dotted with black	Trawl	*Rossia caroli*
1141	21 July	Anchorage S.W. of Santa Lucia Island		15		Line	*Sepia officinalis*
1150	25 July	16°12′ N	24°43′45″ W	3,890	Volcanic muddy sand	Trawl	*Cirroteuthis megaptera*
1182	11 Aug.	14°47′ N	24°31′45″ W	2,478	Muddy black sand	Trawl	*Eledonella diaphana*
1222	24–30 Aug.	Tarrafal Bay (San Antonio) Cape Verde Island		On the beach			*Spirula peroni*
1228	5 Sept.	30°50′ N	25°20′45″ W	Surface		Shrimp net	Oegopsid embryo
1235	8 Sept.	32°34′ N	17°02′45″ W	Surface		Shrimp net	*Leachia cyclura*

EXPEDITION OF 1902

Station No.	Date	Locality		Depth in meters	Nature of bottom	Method of capture	Species caught
		Latitude	Longitude				
1253	21 March	43°25′ N	8°10′15″ E	Surface		Harpoon	Several fragments in the stomach of a Globicéphale
1307	29 July	37°18′ N	20°15′45″ W	Surface		Shrimp net	*Chaunoteuthis mollis*, tentacle of *Mastigoteuthis*.
1308	30 July	37°28′ N	17°49′45″ W	Surface		Shrimp net	*Taonius pavo*
1325	7–8 Aug.	38°39′ N	25°06′15″ W	?		On a trawl line	Tentacle of *Mastigoteuthis*.
1334	13 Aug.	39°30′ N	29°02′15″ W	1,900	Globigerina mud, volcanic sand	Trawl	*Cirroteuthis grimaldii*
1339	14 Aug.	38°41′30″ N	28°45′15″ W	950	Sand and volcanic gravel	Trawl	*Mastigoteuthis* sp.
1340	15 Aug.	Fayal (Azores)					*Octopus vulgaris* bought in the market.
1408	3 Sept.	38°01′30″ N	26°16′45″ E	?		On a sounding line	Tentacle of *Mastigoteuthis*
1409	4 Sept.	39°40′ N	27°14′45″ W	Surface		Shrimp net	*Leachia cyclura*
1440	14 Sept.	47°12′ N	10°29′45″ W	Surface		Long lines	Indeterminate juvenile oegopsid in the stomach of a tuna.
1442	14 Sept.	48°21′ N	9°17′45″ W	152		Beam trawl	*Todaropsis eblanae*

EXPEDITION OF 1903

Station No.	Date	Locality		Depth in meters	Nature of bottom	Method of capture	Species caught
		Latitude	Longitude				
1447	23 July	45°21′ N	2°39′ W	130	Fine sand	Beam trawl	*Illex illecebrosus*
1450	24 July	45°09′ N	3°18′ W	1,804	Sandy mud	Trawl	*Cirroteuthis grimaldii*
1455	25 July	45°13′ N	3°06′ W	358	Sandy mud	Trawl	*Rossia macrosoma, Illex illecebrosus, Todaropsis eblanae*
1475	3 Aug.	44°35′05″ N	1°57′30″ W	140	Sand	Beam trawl	*Sepia elegans, Sepia orbignyana*
1497	11 Aug.	43°31′ N	2°05′5″ W	208	Sandy mud	Beam trawl	*Octopus vulgaris, Rossia macrosoma, Todaropsis eblanae*
1507	14 Aug.	44°34′ N	4°38′30″ W	4,330	Mud	On a bottom line	Fragments in the stomach of *Aphanopus carbo*
1535	1 Sept.	47°46′ N	5°40′ W	132	Sand and broken shells	Beam trawl	*Eledone cirrhosa, Sepia elegans*
1540	4 Sept.	47°16′ N	5°16′ W	140	Sand and shells	Beam trawl	*Eledone cirrhosa, Sepia elegans*
1583	15 Sept.	47°36′ N	7°38′ W	1,490	Muddy sand	Trawl	*Stigmatoteuthis chuni*

EXPEDITION OF 1904

Station No.	Date	Locality		Depth in meters	Nature of bottom	Method of capture	Species caught
		Latitude	Longitude				
1715	1 Aug.	28°04′ N	16°49′30″ W	0 to 1,000		Large Richard net	*Octopus* sp. juvenile, *Pterygioteuthis* sp. juvenile, *Teuthowenia megalops* juvenile, *Hensenioteuthis joubini*
1746	9 Aug.	30°41′ N	17°46′ W	On the bridge			*Stenoteuthis bartrami* juvenile
1749	9 Aug.	30°41′ N	17°46′ W	0 to 2,500		Large Richard net	*Teuthowenia megalops* juvenile
1751b	10 Aug.	30°01′ N	17°24′ W	?		On a sounding line	Tentacle of *Mastigoteuthis*
1763	14–16 Aug.	Las Palmas	(Canaries)			On the beach	*Spirula peroni*
1768	17 Aug.	27°43′ N	18°28′ W	0 to 3,000		Large Richard net	*Tremoctopus violaceus*, *Spirula peroni* young, cranchiid young
1794	25 Aug.	31°46′ N	25° W	0 to 3,000		Large Richard net	*Thelidioteuthis alessandrini* juvenile
1797	26 Aug.	32°18′ N	23°58′ W	0 to 2,000		Large Richard net	*Cranchia scabra*
1799	26 Aug.	32°18′ N	23°58′ W	Surface		Shrimp net	*Leachia cyclura*
1800	26 Aug.	32°18′ N	23°58′ W	0 to 1,000		Large Richard net	*Fusocranchia alpha*
1834	6 Sept.	37°28′ N	25°53′30″ W	0 to 1,000		Large Richard net	*Abraliopsis* sp. juvenile
1849	8 Sept.	36°17′ N	28°53′ W	0 to 3,000		Large Richard net	*Teuthowenia megalops* juvenile
1850	8 Sept.	36°17′ N	28°53′ W	Surface		Shrimp net	*Leachia cyclura*
1856	9 Sept.	36°46′ N	26°41′ W	0 to 3,250		Large Richard net	*Pterygioteuthis* sp. juvenile, *Fusocranchia alpha* juvenile
1872	11 Sept.	37°35′ N	24°40′ W	Surface		Fine-mesh net	*Fusocranchia alpha*
1874	12 Sept.	37°20′ N	21°40′ W	0 to 2,000 (Depth at 3,800)		Large Richard net	*Onychoteuthis banksi* juvenile, *Teuthowenia megalops* juvenile
1875	12 Sept.	37°20′ N	21°40′ W	Surface		Shrimp net	Fragments from the stomach of a turtle
1880	13 Sept.	37° N	18°52′ W	Surface		Shrimp net	*Leachia cyclura*
1886	14 Sept.	36°41′ N	14°03′ W	Surface		Shrimp net	Fragment of arm of a histioteuthid
1893	15 Sept.	36°36′ N	11°49′30″ W	0 to 3,417		On the sounding line	Tentacle of *Mastigoteuthis*

EXPEDITION OF 1905

Station No.	Date	Locality — Latitude	Locality — Longitude	Depth in meters	Nature of bottom	Method of capture	Species caught
1965	30 Mar.	43°18′30″ N	7°16′ E	0 to 2,175	On muddy bottom at 2,225 m	Large Richard net	*Heteroteuthis dispar*
1991	14 Apr.	42°53′ N	8°22′ E	0 to 2,000		Large Richard net	*Galiteuthis armata*
1994	18 Apr.	43°43′20″ N	7°26′30″ E			Beam trawl	*Sepia elegans*, *Sepia orbignyana*, *Todaropsis eblanae*
		In the open sea of Monaco		141	Mud		
2001	20 Apr.	42°58′ N	8°56′30″ E	0 to 1,500		Large Richard net	*Calliteuthis* sp. juvenile
2022	25 July	34°02′ N	12°21′ W	0 to 4,000		Large Richard net	*Chiroteuthis* sp. (2 juveniles)
2058	2 Aug.	29°11′ N	22°01′ W	0 to 500		Large Richard net	*Pterygioteuthis* sp. juvenile
2073	6 Aug.	25°45′ N	33°30′ W	Surface		Fine-mesh net	*Rhynchoteuthis* sp.
2078	7 Aug.	25°58′ N	35°07′ W	Surface		Fine-mesh net	*Rhynchoteuthis* sp.
2108	12–16 Aug.	31°44′30″ N	42°39′ W	3,465	Globigerina mud	Triangular bow net	*Mastigoteuthis magna*
2118	15 Aug.	31°40′30″ N	42°44′30″ W	3,465		Beam trawl	*Melanoteuthis lucens*, *Galiteuthis armata*
2123	12–16 Aug.	31°44′30″ N	42°39′ W	0 to 3,465		On the cable of bow net	Tentacle of *Mastigoteuthis*
2130	17 Aug.	33°03′ N	41°08′ W	0 to 3,000		Large Richard net	*Onychoteuthis banksi* juvenile, *Pterygioteuthis* sp. juvenile, *Ctenopteryx* sp. juvenile, *Mastigoteuthis* sp.
2148	20 Aug.	33°51′ N	34°03′ W	0 to 3,302		On the cable of the sounding line	Tentacle of *Mastigoteuthis*
2153	21 Aug.	35°04′ N	32°11′ W	0 to 2,000		Large Richard net	*Cranchia scabra*
2168	23 Aug.	36°35′ N	27°12′ W	0 to 2,000		Large Richard net	*Melanoteuthis lucens*, *Eledonella diaphana*, *Pterygioteuthis* sp. juvenile
2172	23 Aug.	36°50′ N	26°50′ W	Surface		Fine-mesh net	*Octopodoteuthis sicula* juvenile
2178	28 Aug.	38°03′ N	26°06′ W	Surface		Shrimp net	*Leachia cyclura*
2184	28 Aug.	38°04′45″ N	25°54′ W	Surface		Shrimp net	*Stenoteuthis bartrami*
2186	29 Aug.	38°04′ N	26°07′ W	Surface		Shrimp net	*Leachia cyclura*
2187	29 Aug.	38°04′ N	26°07′ W	0 to 2,500		Large Richard net	*Teuthowenia megalops*

EXPEDITION OF 1906

Station No.	Date	Locality		Depth in meters	Nature of bottom	Method of capture	Species caught
		Latitude	Longitude				
2350b	31 May	40°10′ N	11°29′ W	0 to 2,600		On the sounding line	Tentacle of *Mastigoteuthis*

EXPEDITION OF 1908

Station No.	Date	Locality		Depth in meters	Nature of bottom	Method of capture	Species caught
		Latitude	Longitude				
2688	10 July	41°40′ N	4° E	0 to 2,025		Large Richard net	*Rhynchoteuthis*, embryo of *Spirula*
2689	10 July	41°40′ N	4° E	Surface		Harpoon	Fragments from the stomach of a Globicéphale
2695	12 July	38°35′ N	3°06′ E	0 to 2,595		Large Richard net	*Onychoteuthis banksi* (2 juveniles), *Thelidioteuthis alessandrini* juvenile, *Benthoteuthis megalops* juvenile.
2704	14 July	36°18′ N	2°31′ W	0 to 1,665		Large Richard net	*Spirula* embryo
2714	17 July	35°56′ N	8° W	0 to 1,400		Large Richard net	*Eledonella* sp. juvenile
2786	25 Aug.–8 Sept.	Thamshavn near Trondheim		17 to 20		Trammel net	*Rossia palpebrosa*
2792	11–12 Sept.	Askevold (Norway)		20		Trammel net	*Elodone cirrhosa*

EXPEDITION OF 1909

Station No.	Date	Locality		Depth in meters	Nature of bottom	Method of capture	Species caught
		Latitude	Longitude				
2876	8 Aug.	43°04′30″ N	19°42′ W	0 to 1,000		Large Richard net	*Benthoteuthis megalops* juvenile
2902	27 Aug.	36°17′ N	1°58′ W	0 to 1,800		Large Richard net	*Abraliopsis* sp. (3 juveniles), *Enoploion eustictum* (2 juveniles)
2905	27 Aug.	36°21′30″ N	1°25′ W	0 to 1,200		Large Richard net	*Tracheloteuthis* sp. *Rhynchoteuthis* sp.
2916	29 Aug.	38°20′ N	1°31′ E	0 to 1,000		Large Richard net	*Thelidioteuthis alessandrini* juvenile
2929	7 Sept.	39°40′ N	5°56′ E	Surface		Shrimp net	*Ommatostrephes sagittatus*

EXPEDITION OF 1910

Station No.	Date	Locality		Depth in meters	Nature of bottom	Method of capture	Species caught
		Latitude	Longitude				
2986	17 Aug.	45°07′30″ N	7°06′ W	4,870	Globigerina mud	Trawl	*Cirroteuthis umbellata*
2994	19 Aug.	44°08′ N	10°44′ W	5,000	Globigerina mud	Trawl	*Cirroteuthis umbellata*
3008	23 Aug.	43°12′ N	10°01′ W	Surface		Eel spear	*Ommatostrephes sagittatus* (under the electric light)
3011	24 Aug.	42°26′30″ N	9°29′ W	1,680		On a bottom line	Fragments from the stomach of a dogfish (*Centroscymnus coelolepis*)
3026	3 Sept.	36°31′ N	11°33′30″ W	85		Line	*Octopus vulgaris* from the stomach of *Serranus atricauda*
3028	4 Sept.	36°54′30″ N	11°49′ W	0 to 5,100		Bourée net	*Eledonella diaphana*
3031	5 Sept.	37°10′ N	11°48′ W	Surface		Shrimp net	*Alloposus mollis*
3039	8 Sept.	36°05′30″ N	9°00′30″ W	0 to 3,660		Bourée net	*Eledonella diaphana, Abraliopsis pfefferi, Doratospsis* sp., *Teuthowenia megalops, Bathothauma lyromma* juvenile
3045	9 Sept.	36°13′30″ N	8° W	0 to 1,400		Bourée net	*Galiteuthis armata, Pterygioteuthis giardi*
3047	9 Sept.	36°14′ N	8°06′ W	1401		On a bottom line	Fragments from the stomach of a dogfish (*Centrophorus calceus*)
3050	12 Sept.	36°34′ N	3°07′ W	Surface		Harpoon	Fragments from the stomach of a cetacean (*Globicephalus melas*)
3053	13 Sept.	37°20′ N	0°07′ W	Surface		Fine-mesh net	*Rhynchoteuthis* sp.

91 # Bibliography

1. Appellöf, A. 1890. Teuthologische Beiträge. I. *Ctenopteryx*, n. g., *Verania sicula* Krohn, *Calliteuthis* Verrill, Bergens Museums Aarsberetning for 1889. Bergen.
2. Appellöf, A. 1890. Teuthologische Beiträge. II. *Chaunoteuthis* n. g. *oegopsidarum* s., 29 p., 4 pl. Bergen.
3. Ball, R. 1841–1842. On a Loligo found on the shore of Dublin Bay. *Proc. Royal Irish Acad.*, I. 1841, II. 1842.
4. Cantraine, F. 1840. Malacologie méditerranéenne et littorale. *Nouv. Mém. Acad. des Sciences et Belles Letters de Belgique*, Bruxelles, XIII, p. 173.
5. Carus, V. 1889–1893. Prodromus Faunae Mediterraneae. Cephalopoda: II, Pars II. Stuttgart.
6. Chun, Carl. 1910. *Cirrothauma* ein Blinder Cephalopod. Leipzig.
7. Chun, Carl. 1906. System der Cranchien. *Zoolog. Anzeiger*. Leipzig.
8. Chun, Carl. 1908. Uber Cephalopoden der deutschen Tiefsee-Expedition. *Zool. Anzeiger*. XXXIII, Leipzig.
9. Chun, Carl. 1910. Wissenschaftliche Ergebnisse der deutschen Tiefsee-Expedition. Die Cephalopoden, Teil I. Oegopsida, Leipzig.
10. Delle Chiaje, S. 1829. Memorie sulla storia e notomia degli animali senza vertebre del regno di Napoli, Vol. 4. Napoli.
11. Eschricht, D.F. 1836. *Cirroteuthis mulleri*, eine neue Gattung der Cephalopoden bildend. *Nova Acta Acad. Caes. Leopold. Car. Naturae. Curiosorum,* XVIII, pt. 11, p. 627. Dresden.
12. Férussac, J.-B.-L. 1834. Histoire naturelle generale et particulière des Mollusques, etc. Paris.
13. Fischer, Paul. 1883. Note préliminaire sur une nouvelle espèce du genre *Cirroteuthis*. *Journ. de Conch.*, Vol. 31. Paris.
14. Fischer, H. 1896. Note préliminaire sur le *Pterygioteuthis giardi*, Cephalopode nouveau recueilli dans le cours de l'expédition scientifique du TALISMAN (1883). *Journal de Conchyliologie*, Vol. 43, p. 205. Paris.
15. Fischer, H. and L. Joubin. 1906. Notes sur les Céphalopodes capturés au cours des expéditions du TRAVAILLEUR et du TALISMAN. *Bull. Museum de Paris,* XII.
16. Fischer, H. and L. Joubin. 1907. Expéditions scientifiques du *TRAVAILLEUR* et du *TALISMAN. Céphalopodes*, Vol. VIII. Paris.
17. Gillivray, Mc. W. 1843. *A history of the Molluscous and Cirripedial animals of Scotland as found in the North-Eastern district*, London.

92 18. Girard, Albert, A. 1890. Révision des Céphalopodes du Museum de Lisbonne. *Jornal de Sciencias Math. Phys. et Nat., publicado sob os auspicios da Academia das Sciencias II,* Vol. 1, No. IV. Lisboa.

19. Girard, Albert, A. 1890. Révision des Céphalopodes du Museum de Lisbonne. Additions. *Jornal de Sciencias Math., Phys. et. Nat., publicado sob os auspicios da Academia real das Sciencias II*, Vol. 1, No. V. Lisboa.
20. Girard, Albert. A. 1892. Les Céphalopodes des Iles Acores et de l'Ile de Madére. *Jornal de Sciencias Math., Phys. et. Nat., publicado sob os auspicios da Academia real das Sciencias* (II), Vol. 2, No. VII. Lisboa.
21. Gray, J.-E. 1849. Catalogue of the Mollusca in the collection of the British Museum, Part I, *Cephalopoda antepedia*, London.
22. Hoyle, W.-E. 1884. On *Loligopsis* and some other genera. *Proc. Roy. Phys. Soc.*, Vol. 8. Edinburgh.

23. Hoyle, W.-E. 1885. Diagnoses of new species of Cephalopoda collected during the cruise of the *H.M.S. CHALLENGER*. I. The Octopoda. *Ann. and Mag. of Nat. Hist.*, (V), Vol. 15, London II. The Decapoda. Ibid. (V), Vol. 16.

24. Hoyle, W.-E. 1886. A catalogue of recent Cephalopods. *Proc. Roy. Phys. Soc.*, Vol. 9. Edinburg. First Supplement 1897, Second Supplement 1906.

25. Hoyle, W.-E. 1886. Report on the scientific results of the voyage of the *H.M.S. Challenger* during the years 1873–1876. Zoology, Report on the Cephalopods, Vol. 16, part 44. Edinburgh.

26. Hoyle, W.-E. 1901. On the generic names *Octopus*, *Eledone* and *Histopsis*. *Mem. Manchester lit. et. phil. Soc.*, Vol. 45, No. 9.

27. Hoyle, W.-E. 1904. Reports on the dredging operations, carried on by the *ALBATROSS*. *Bull. Mus. Compar. Zool.*, Vol. 43. Cambridge.

27b. Huxley, Th. H. and Paul Pelseneer. 1895. Observations sur la *Spirula*. *Bulletin scientifique de la France et de la Belgique*, XXXI.

28. Jatta, Giuseppe. 1896. Cefalopodi viventi nel golfo di Napoli, in: Fauna und Flora des Golfes von Neapel, Monog, 23, Cefalopodi (Systematica), Berlin.

29. Joubin, Louis. 1893. Recherches sur l'appareil lumineux d'un Céphalopode (*Histioteuthis rüppelli* Vérany). *Bull. Soc. Scient. et Médic. de l'Ouest*, Vol. 2. Rennes.

30. Joubin, Louis, 1893. Note complémentaire sur l'appareil lumineux d'un Céphalopode (*Histioteuthis rüppelli* Vérany). *Bull. Soc. Scient. et Médic. de l'Ouest*, Vol. 2. Rennes.

31. Joubin, Louis. 1894. Nouvelles recherches sur l'appareil lumineux des Céphalopodes du genre *Histioteuthis*. *Bull. Soc. Scient. et Médic. de l'Ouest*, Vol. 3. Rennes.

32. Joubin, Louis. 1895. Note sur les appareils photogènes cutanés de deux Céphalopodes: *Histiopsis atlantica* Hoyle et *Abralia oweni* (Vérany) Hoyle. *Mém. Soc. Zool. de France*, Vol. 8. Paris.

33. Joubin, Louis. 1895. Contribution à l'étude des Céphalopodes de l'Atlantique Nord, Res. Camp. Scient. accomplies sur son yacht par S. A. S. Albert 1er, Prince Souverain de Monaco. fasc. XII, Monaco.

34. Joubin, Louis. 1896. Observations sur divers Céphalopodes. Note I: *Abraliopsis pfefferi* nov. gen. et sp. *Bulletin Soc. Scient. et Médic de l'Ouest*, Vol. 5, No. 1, pp. 19–35.

93 35. Joubin, Louis. 1898. Note sur une nouvelle famille de Céphalopodes. *Ann. des Sc. nat. Zoologie*, VI. Paris.

36. Joubin, Louis. 1898. Observations sur divers Céphalopodes. Quatrième note: *Grimalditeuthis richardi*. *Bull. Soc. Zool. de France*, Vol. 23. Paris.

37. Joubin, Louis. 1900. Céphalopodes provenant des Campagnes de la *PRINCESSE-ALICE* (1891–1897), Résultats des campagnes scientifiques accomplies sur son yacht par S. A. S. Albert 1er, Prince Souverain de Monaco. Fasc. XVII, with 15 plates, Monaco.

38. Joubin, Louis. 1902. Observations sur divers Céphalopodes. Sixième note: Sur une nouvelle espèce du genre *Rossia*. *Bulletin de la Societé Zoologique de France*, Vol. 27, pp. 138–143. Paris.

39. Joubin, Louis. 1902. Révision des Sepiolidae. *Mem. Soc. Zool. de France*, XV, pp. 80–145.

40. Joubin, Louis. 1902. Observations sur divers Céphalopodes. Note septième: *Heteroteuthis weberi* n. sp. *Bull. Soc. Scientifique et Médicale de l'Ouest*, Vol. XI, pp. 401–404. Rennes.

41. Joubin, Louis. 1903. Sur quelques Céphalopodes recueillis pendant les dernières campagnes de S. A. S. le Prince de Monaco (1901–1902). *Comptes rendus de l'Académie des Sciences*, Vol. 136, pp. 100–102. Paris.

42. Joubin, Louis. 1910. Observations sur une jeune *Spirula*. *Bull. Institut Océanographique*, No. 165, 29 mars. Monaco.

43. Joubin, Louis. 1912. Études préliminaires sur les Céphalopodes recueillis au cours des croisières de S. A. S. le Prince de Monaco. 1re note: *Melanoteuthis lucens* nov. gen. et sp. *Bulletin de l'Institut Océanographique*, No. 220, 10 janvier. Monaco.

44. Joubin, Louis. 1912. Études préliminaires sur les Céphalopodes recueillis au cours des croisières de S. A. S. le Prince de Monaco. 2e note: *Cirroteuthis grimaldii* nov. sp. *Bulletin de l'Institut Océanographique*, No. 226, 25 mars. Monaco.

45. Joubin, Louis. 1913. Études préliminaires sur les Céphalopodes recueillis au cours des croisières de S. A. S. le Prince de Monaco. Troisième note: *Mastigoteuthis magna*. *Bulletin de l'Institut Océanographique*, No. 275. Monaco.

46. Krohn, A. 1845–1847. Uber einen neuen Cephalopoden (*Octopodoteuthis*). *Arch. Naturg*, XI, p. 45, 1845, suppl. XIII, 1847, p. 36.

47. Lamarck, Le Citoyen. 1799. Sur les genres de la Sèche, du Calmar et du Pouple, vulgairement nommés Polypes de mer. *Mém. Soc. Hist. Nat.*, Vol. 1. Paris.

48. Lamarck, Le Chevalier de. 1822. Histoire naturelle des Animaux sans vertèbres, présentant les charactères généraux et particuliers de ces animaux, leur distribution, leur classes, leurs familles, leurs genres, et la citation des principales espèces qui s'y rapportent. Vol. 6, Paris, 1814; Vol. 9, Paris.

49. Leach, W.-E. 1817. Zoological miscellany, being description of new or interesting animals, illustrated with colored figures drawn from nature by R.P. Nodder, Vol. 3. London.

50. Leach, W.-E. 1818. Sur plusieurs espèces nouvelles de la classe des Céphalopodes et sur une nouvelle distribution systématique des ordres, familles et genres de cette classe. *Journ. de Phys., de Chimie et d'Hist. Nat.*, Vol. 86. Paris.

94 51. Leach, W.-E. 1852. Molluscorum Brittaniae synopsis. A synopsis of the Mollusca of Great Britain, arranged according to their natural affinities and anatomical structure. London.

52. Lesueur. 1821. Description de plusieurs espèces nouvelles de Seiches. *Journ. of the Nat. Acad. of Science*, Vol. 2. Philadelphia.

53. Lichtenstein, H. 1818–1819. Von den Sepien mit Krallen. *Abhdl. Berlin Acad.*, Vol. 7. Berlin.

54. Linné, Carolus A. 1761. Fauna Suecica.

55. Linné, Carolus A. 1766–1767. Systema Naturae, per regna tria naturae secundum classes, ordines, genera, species, cum characteribus, differentiis, synonymis, locis; editio duodecima reformata, tomus 1, Holmiae.

56. Lovén, S. 1845. Om nordiska Cephalopoder. *Ofversigt Svenska. Vet. Akad. Forh.*, IV. Stockholm.

57. Lönneberg, Einar. 1895–1898. On the Cephalopods collected by a Swedish expedition to Tierra del Fuego. *Wiss. Beob. Exped. nach den Magellansländern 1895–1897*. II. Stockholm, 1898.

58. Lütken, 1881–1882. Dyreriget. Kjøbenhavn.

59. Mörch, O. 1850. Om *Cranchia megalops* Prosch. *Vid. Med. Nat. For.* Kjøbenhavn.

60. Norman, A.-M. 1890. Revision of British Mollusca. *Ann. and Mag. of Nat. Hist.*, [VI], Vol. 5. London.

61. Orbigny, A. d'.1826. Tableau méthodique de la classe des Céphalopodes. *Annales des Sciences naturelles*, Paris.

62. Orbigny, A. d'. 1836–1843. Voyage dans l' Amérique Méridionale, exécuté dans le cours des années 1826–1833. *Mollusques*, Vol. 2, partie 3. Paris.

63. Orbigny, A. d'and D. de Férussac. 1835–1848. Histoire naturelle générale et particulière des Céphalopodes acétablifères vivants et fossiles. Paris.

64. Orbigny, A. d'. 1846. Paléntologie universelle des Coquilles et des Mollusques. Paris.

65. Owen, R. 1835. J. Ross appendix to the narrative of a second voyage in search of a North West passage, 1829–1833, London.

66. Owen, R. 1836. Description of some new and rare Cephalopods. *Trans. Zool. Soc.*, Vol. 2. London.

67. Owen, R. 1881. Description of some new and rare Cephalopods. *Trans. Zool. Soc.*, Vol. 11. London.

68. Pfeffer, G. 1884. Die Cephalopoden des Hamburger naturhistorischen Museums. *Abhandl. aus dem Geb. der Nat. heraus. v. naturwissensch. Verein*, Vol. 8. Hamburg.

69. Pfeffer, G. 1900. Synopsis der oegopsiden Cephalopoden. *Mittheil. aus dem Naturhistorischen Museum*, XVII, Hamburg.

70. Pfeffer, G. 1912. Die Cephalopoden der Plankton-Expedition. Ergebnisse der Plankton-Expedition der Humboldt Stiftung. Kiel und Leipzig.

71. Prosch, V. 1847. Nogle nye Cephalopoder beskrevne og anatomisk undersögt. *Overs. Danske vid. Selskabs. Forh. I,* Kjøbenhavn.

95 72. Rafinesque, Sch. 1814. Précis des découvertes somiologiques entre 1800 et 1801, ou choix des principales découvertes en zoologie et en botanique. Palerme.

73. Reinhardt, J.T. and V. Prosch. 1846. Om *Sciadephorus mülleri* Eschricht. *K. Danske Vid. Selskab*. Skrifter, XII, p. 165. Kjøbenhavn.

74. Rochebrune, T. de. 1884. Etude Monographique de la famille des Loligopsidae. *Bull. de la Soc. Philomatique*, VII, Vol. 8. Paris.

75. Rüppell, E. 1847. Intorno ad alcuni Cefalopodi del mare di Messina. Giorn. del gabinetto di Messina, XXVII.

76. Schneider, J.G. 1784. Characteristik des ganzen Geschlechts und der einzelnen Arten von Blackfischen. *Samml. vermischt Abhandl. z. Aufklärung d. Zool.* Berlin.

77. Steenstrup, J. 1861. Cephalopodum, qui in museis Hafniensibus inveniuntur, Pelagicorum synopseos particula prima. *Overs. Dan. videns. Selsk.* Kjøbenhavn.

78. Steenstrup, J. 1881. En ny Bloecksprutterslоegt: *Tracheloteuthis. Vidensk. meddel. Natur. Forh.* Kjøbenhavn.

79. Steenstrup, J. 1887. Notae Teutologicae 6. Species generis *Sepiolae* maris Mediterranei. *Overs. K.D. vidensk. Selsk. Forh.* Kjøbenhavn.

80. Troschel, H. 1857. Bemerkungen uber die Cephalopoden von Messina. *Arch. f. naturges.*, 23.

81. Tryon, G.W. 1879–1889. Manual of Conchology, structural and systematic, with illustrations of the species. Philadelphia.

82. Vérany, J.B. 1839. Mémoire sur six nouvelles espèces de Céphalopodes trouvées dans la Méditterranée. *Mem. R. Accad. d. Sc.*, (II), Vol. 2. Torino.

83. Vérany, J.B. 1851. Méllusques méditerranéens observés decrits, figurés et chromolithographiés d'après nature sur des modèles vivants, 1re part, Céphalopodes de la Mediterranée. Gênes.

84. Verrill, A.E. 1873. *Octopus bairdi* Verrill. *Amer. Journ. Science*, V, p. 5.

85. Verrill, A.E. 1880. Synopsis of the Cephalopods of the North Eastern coast of America. *Amer. Journ. Sc. and Arts*, Vol. 20. New-Haven.

86. Verrill, A.E. 1881. Report on the Cephalopods and on some additional species dredged by the U.S. Fish commission steamer FISH-HAWK, during the season of 1880, *Bull. Mus. Comp. Zoöl. at Harvard College*, Vol. 8, No. 5. Cambridge.

87. Verrill, A.E. 1882. Report on the Cephalopods of the north-eastern coast of America. Annual report of the commissioner of Fisheries for 1879. Washington.

88. Verrill, A.E. 1881–1885. Second catalogue of Mollusca recently added to the fauna of the New England coast and the adjacent parts of the Altantic, consisting mostly of deep sea species, with notes on others previously recorded. *Trans. Connect. Acad.*, Vol. 6. New-Haven.

89. Verrill, A.E. 1885. Third catalogue of Mollusca, recently added to the fauna of the New England coast and adjacent parts of the Atlantic. *Trans. Connect. Acad.*, Vol. 7. New-Haven.

Plates

PLATE I

Figs. 1–4. *Cirroteuthis megaptera* Verrill.

Fig. 1. Dorsal view. × 2.5.

Fig. 2. Ventral view. × 2.5.

Fig. 3. Profile of the animal. × 2.5.

Fig. 4. Interior of the web. × 2.5.

PLATE I

PLATE II

PLATE II

Cirroteuthis grimaldii Joubin.
Dorsal view of the animal. Natural size.

PLATE III

Cirroteuthis grimaldii Joubin.
Ventral view of the animal. Natural size.

PLATE IV

Figs. 1–4. *Cirroteuthis grimaldii* Joubin.

Fig. 1. Animal seen in profile, reduced to approx. one-third. Left side.

Fig. 2. Animal seen in profile, reduced to approx. one-third. Right side.

Fig. 3. Interior of the web. Photograph reduced to approx. one-third.

Fig. 4. Close-up view of the arms and suckers. Natural size.

PLATE V

Melanoteuthis lucens Joubin.
Ventral view of the animal. Approx. × 5

The figure in this plate was drawn by Miss Vesque based on the watercolor sketch made by Mr. L. Tinayre at the moment of its capture.

PLATE V

PLATE VI

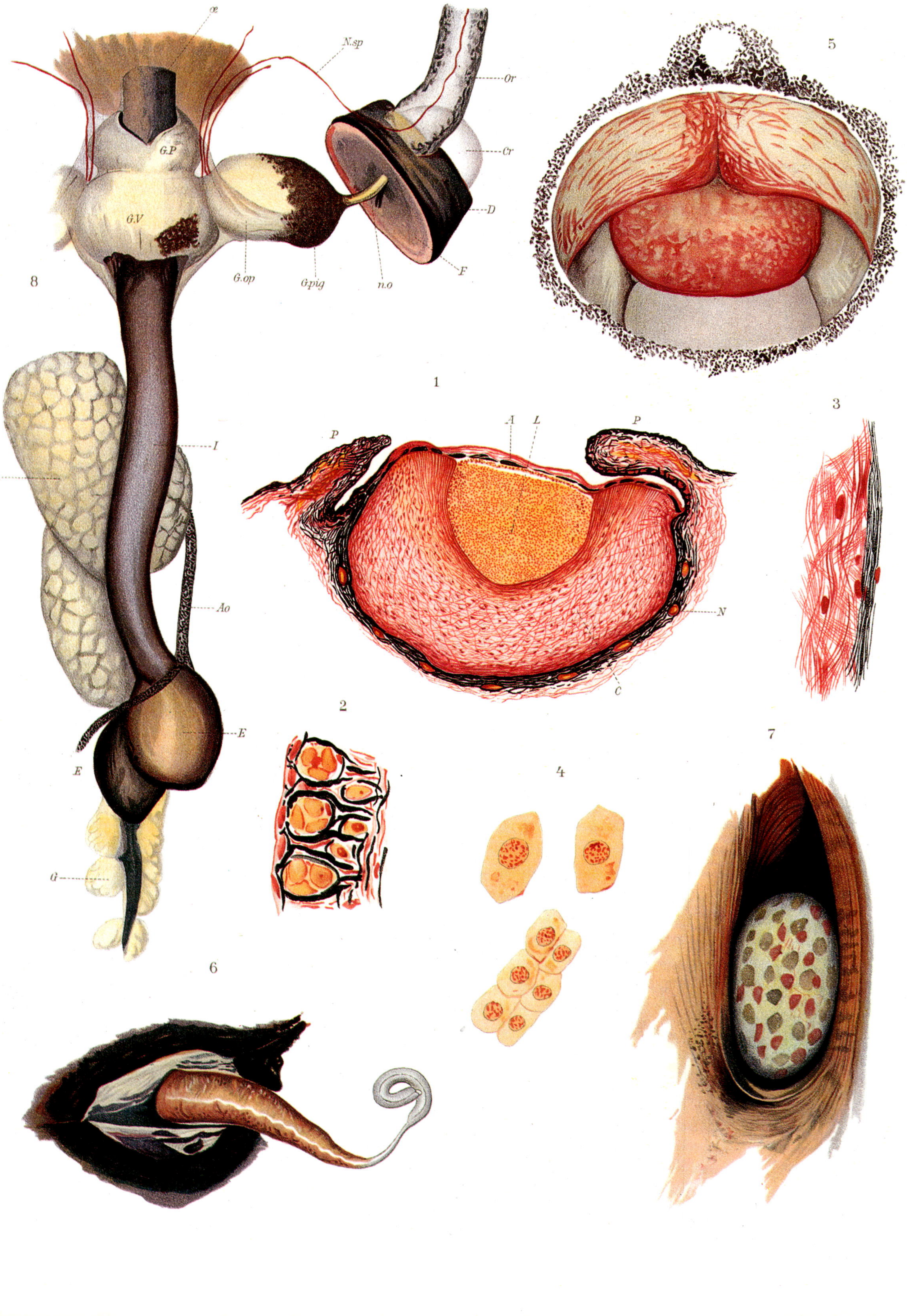

PLATE VI

Figs. 1–8. *Melanoteuthis lucens* Joubin.

Fig. 1. Traverse section through the middle of the luminous organs. × 30. (A) cornea; (C) connective tissue layer; (L.) central granular tissue; (N.) black pigment layer; (P.) lidfold.

Fig. 2. Section of the lidfold showing the bundles of muscle fibres entwined with pigmented cells. × 300.

Fig. 3. Section of the fibrous region, stained red, at one of its points of contact with the black pigmented region. × 770.

Fig. 4. Cells of the central granular region, showing the large nucleus—granular and deeply stained. × 850.

Fig. 5. Interior view of the funnel showing the disposition of the valve deeply stained with carmine. Approx. × 10.

Fig. 6. Brachial organ in the interior of the umbrella seen in natural color. × $4\frac{1}{2}$.

Fig. 7. Luminous organ in its natural colors. Approx. × 10.

Fig. 8. Dissection of a part of the organs of *Melanoteuthis lucens*, ventral side. × 15.

(Oe) esophagus; (Nsp) special nerve of the brachial organ; (Or) brachial organ; (Cr) lens; (D) eye, lateral wall; (F) eye, flat bottom of ocular chamber; (no) optic nerve; (Gp) pedal ganglion; (GV) visceral ganglion; (I) intestine; (Gh) hepatic gland; (Ao) aorta; (E) stomach; (G) gland (renal ?); (G. op) optic ganglion; (G. pig) pigmented part of optic ganglion.

PLATE VII

Figs. 1–3. *Melanoteuthis lucens* Joubin.

Fig. 1. Left luminous organ. Approx. × 4.

Fig. 2. Right luminous organ. Approx. × 4.

Fig. 3. Dorsal region of the fins showing two luminous organs at their base.

Figs. 4–5. *Octopus arcticus* Prosch.

Fig. 4. Dorsal view of the male. Natural size.

Fig. 5. Hectocotylus of the same. Approx. × 3.

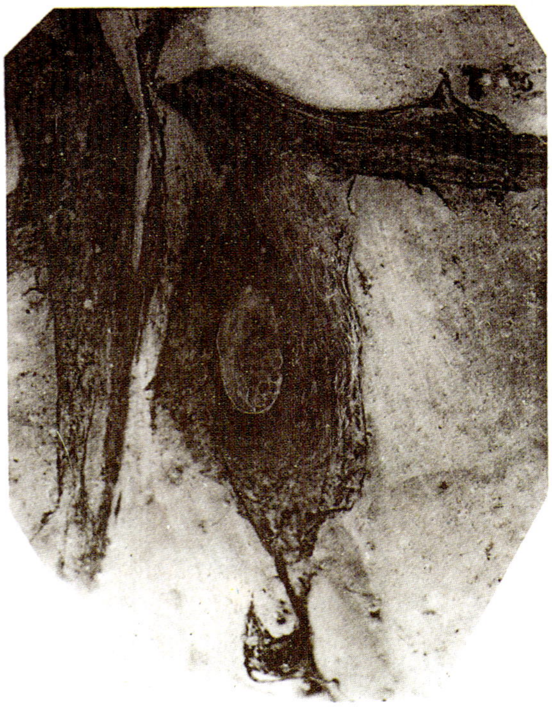

1

2

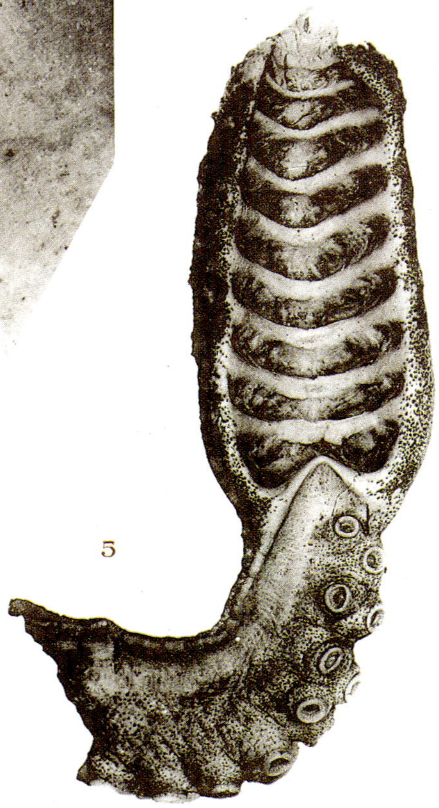

5

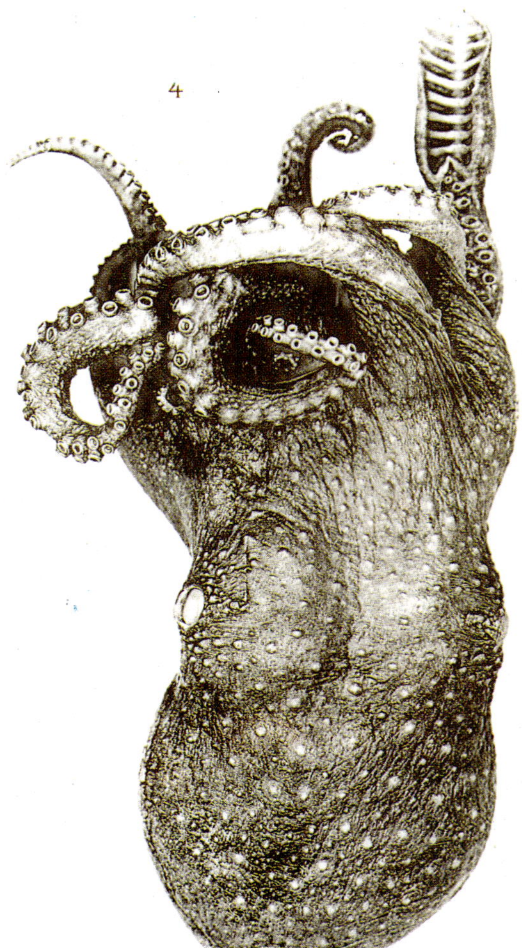

4

3

PLATE VIII

Figs. 1–4. *Rossia caroli* Joubin.

Fig. 1. Dorsal view of the first specimen. Natural size.

Fig. 2. Ventral view of the first specimen. Natural size.

Fig. 3. Dorsal view of the second specimen. Slightly enlarged.

Fig. 4. Ventral view of the second specimen. Slightly enlarged.

Figs. 5–6. *Rossia glaucopis* Lovén.

Fig. 5. Ventral view of the animal. Approx. × 3.

Fig. 6. Dorsal view of the animal. Several spermatophores adhering below the left eye.

Fig. 7. *Todaropsis eblanae* Ball.
Two hectocotylized arms. Slightly enlarged.

Figs. 8–9. *Eledonella* sp. Juvenile.

Fig. 8. Dorsal view. Approx. × 12.

Fig. 9. Ventral side of the head.

Fig. 10 *Bentholeuthis megalops* Verrill.
Enlarged embryo.

PLATE IX

Figs. 1–5. *Spirula* sp. Young.

Fig. 1. Dorsal view of the animal. Approx. × 20.

Fig. 2. Right side of the animal. Approx. × 20.

Fig. 3. Three-fourths of the ventral view of the animal. Approx. × 20.

Fig. 4. Left side of the animal. Approx. × 17.

Fig. 5. Ventral view of the animal. Approx. × 20.

Figs. 6–7. *Mastigoteuthis magna* Joubin.

Fig. 6. Side view of the arm sucker. × 19.

Fig. 7. Front view of the arm sucker. × 22.

Figs. 8–9. Pelagic embryo of an unknown decapod caught at station 1228.

Fig. 8. Ventral side. × 20.

Fig. 9. Dorsal side. × 20.

Fig. 10. *Calliteuthis* sp. Juvenile.

Right side of a juvenile specimen, measuring 5 mm in length, caught at station 2001. × 21.

PLATE IX

PLATE X

PLATE X

Mastigoteuthis magna Joubin.
Ventral view of the animal. Natural size.

The figure in this plate is drawn by Miss Vesque based on the color sketch made by Mr. L. Tinayre at the moment of capture.

PLATE XI

Figs. 1–5. *Mastigoteuthis magna* Joubin.

Fig. 1. Right side of the head. Natural size.

Fig. 2. Interior of the brachial crown showing the disposition of the lip and suckers. Natural size.

Fig. 3. Adhesive cephalic cartilage. × 2.

Fig. 4. Tentacle. Natural size.

Fig. 5. Tentacular sucker. Approx. × 300.

Figs. 6–9. Embryos of *Spirula*.

Fig. 6. Dorsal side. Approx. × 34.

Fig. 7. Same; ventral side.

Fig. 8. Side view of the second specimen. Approx. × 30.

Fig. 9. Same; close-up of the head. Approx. × 40.

1
2
3
4
5
6
7
8
9

PLATE XII

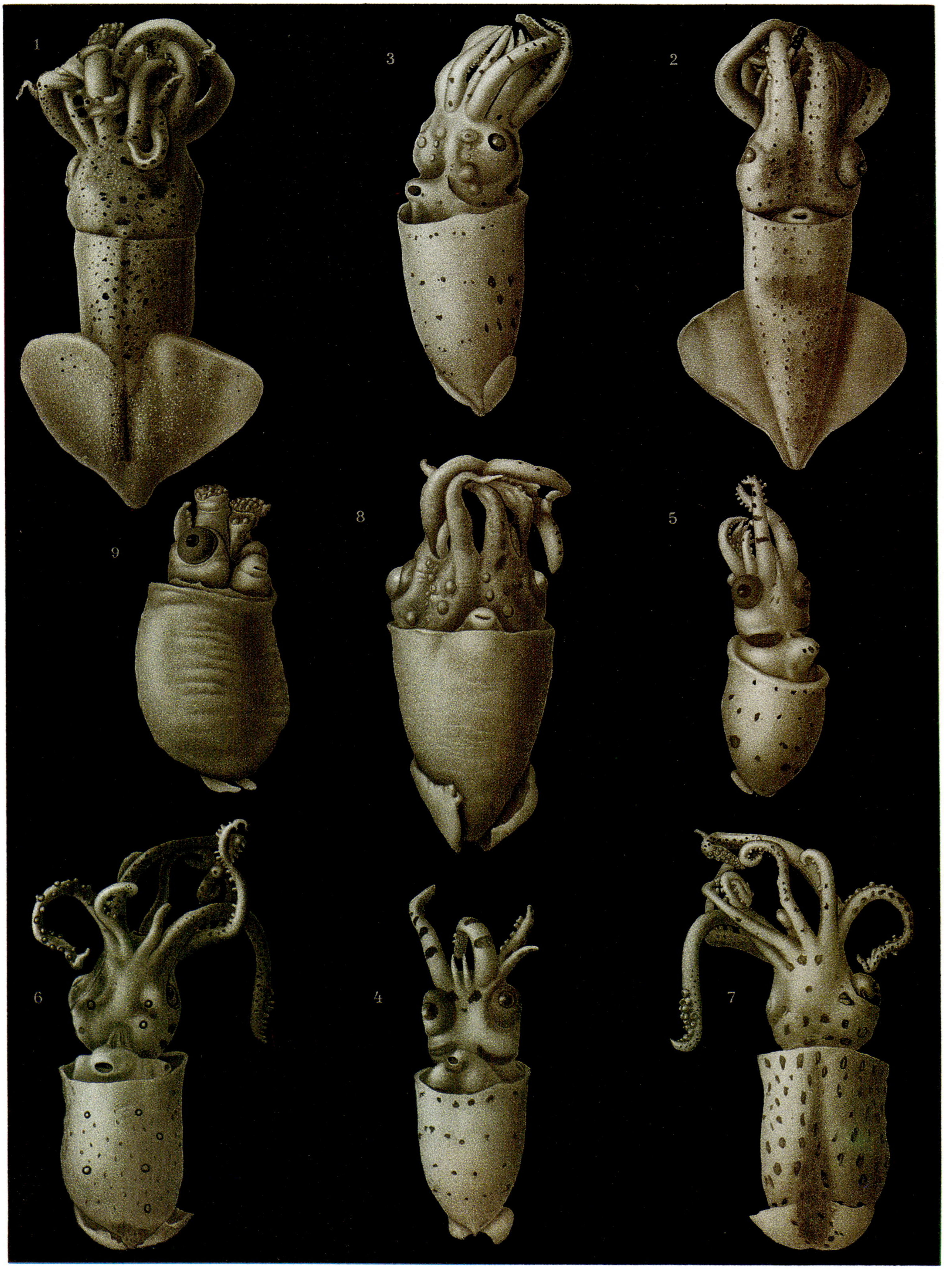

PLATE XII

Figs. 1–2. *Abraliopsis pfefferi* Joubin.

Fig. 1. Dorsal side of a juvenile specimen caught at station 3039. × 6.

Fig. 2. Same, ventral side. × 5.5.

Figs. 3–5. *Abraliopsis* sp. Juvenile.

Fig. 3. Ventral side of the largest of three specimens caught at station 2902. × 8.

Fig. 4. Ventral side of the medium size specimen caught at station 2902. × 11.

Fig. 5. Ventral side of the smallest specimen caught at station 2902. × 12.

Figs. 6–7. *Thelidioteuthis alessandrini* (Vérany).

Fig. 6. Ventral side of a young specimen caught at station 2916. × 5.3.

Fig. 7. Same, dorsal side. × 5.3.

Fig. 8. *Pyroteuthis* sp. Juvenile.
Ventral view of an animal caught at station 2130. × 13.

Fig. 9. *Benthoteuthis* sp.
Ventral side of a very young specimen caught at station 2695. × 20.

PLATE XIII

Fig. 1. *Ctenopteryx* sp.
Dorsal view of a young specimen caught at station 2130. × 12.

Fig. 2. *Tracheloteuthis* sp.
Dorsal view of a young specimen caught at station 2905. × 8.

Fig. 3. *Stigmatoteuthis chuni* Pfeffer.
Ventral view of a young specimen, almost adult, caught at station 1583. Approx. 2.25.

Fig. 4. *Benthoteuthis megalops* Verrill.
Ventral view of a young specimen caught at station 2876. × 10.

Figs. 5–6. *Enoploion eustictum* Pfeffer.

Fig. 5. Young specimen, measuring 6.5 mm, caught at station 2902. × 13.

Fig. 6. Side view of a young specimen, measuring 5.6 mm, caught at station 2902. × 15.

Figs. 7–8. *Thelidioteuthis alessandrini* (Vérany).

Fig. 7. Ventral side of young specimen caught at station 1794. × 11.

Fig. 8. Same specimen, dorsal side. × 8.

7
1
5
6
8
4
3
2

2
5
3
6
1
7
8
4
9

PLATE XIV

Fig. 1. *Galiteuthis armata* Joubin.
Ventral view of a very young specimen caught at station 1991. × 8.

Figs. 2–3. *Hensenioteuthis joubini* Pfeffer.

Fig. 2. Dorsal view of a very young specimen. × 8.

Fig. 3. Same, ventral view. × 8.

Fig. 4. *Bathothauma lyromma* Chun.
Ventral view of a very young specimen caught at station 3039. × 7.

Figs. 5–9. *Fusocranchia alpha* Joubin.

Fig. 5. Dorsal view of a young specimen caught at station 1800. × 8.25.

Fig. 6. Same specimen ventral view. × 11.7.

Fig. 7. Dorsal side of the head of the same specimen. × 11.7.

Fig. 8. Ventral view of a young larva of this species caught at station 1856. × 14.

Fig. 9. Ventral view of another very young larva of this species caught at station 1872. × 20.

PLATE XV

Fig. 1. *Cirroteuthis grimaldii* Joubin.
Diagram showing the fin disposition, its muscles and internal cartilage. Slightly enlarged.

Figs. 2–7. *Melanoteuthis lucens* Joubin.

Fig. 2. Dorsal side of the body showing its shape, position of fins and luminous organs. Nearly natural size.

Fig. 3. Ventral side of the body showing the opening of the mantle slit.

Fig. 4. Structure of the arm viewed from the inner side of the umbrella. Natural size.

Fig. 5. Fragment of the free part of an arm. Approx. × 5.

Fig. 6. Free tip of the umbrella organ. Approx. × 14.

Fig. 7. Interior of the funnel showing its anterior aperture. Approx. × 8.

Fig. 8. *Eledonella* sp. Juvenile.

Anatomical details seen through the transparent mantle.
(B.) buccal bulb; (C.) cerebral ganglion; (I.) digestive canal; (N.B.) brachial nerve; (G.O.) optic ganglion; (G.E.) stellate pallial ganglion; (O) eye; (N.P.) pallial nerve. × 11.5.

PLATE XV

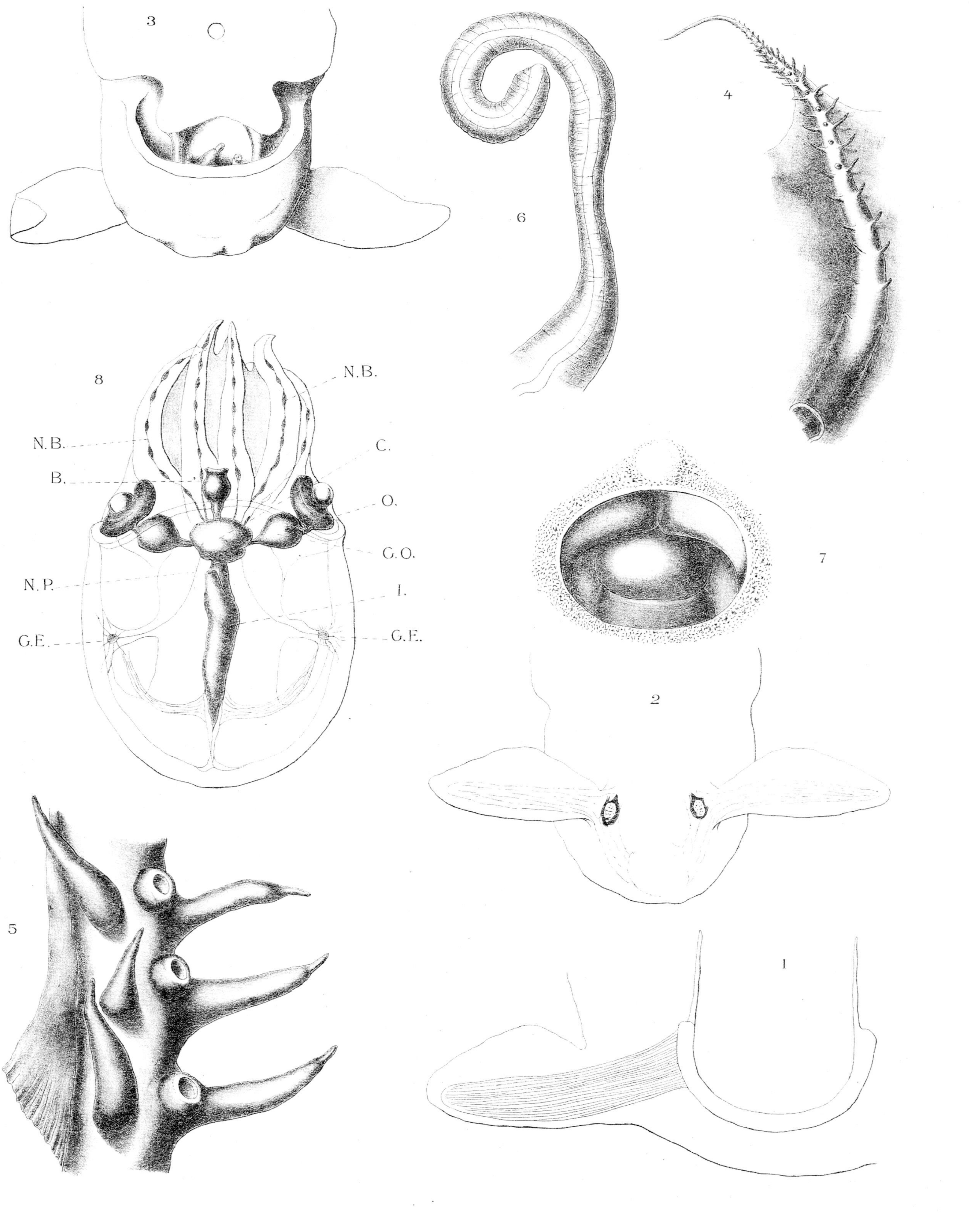

PLATE XVI

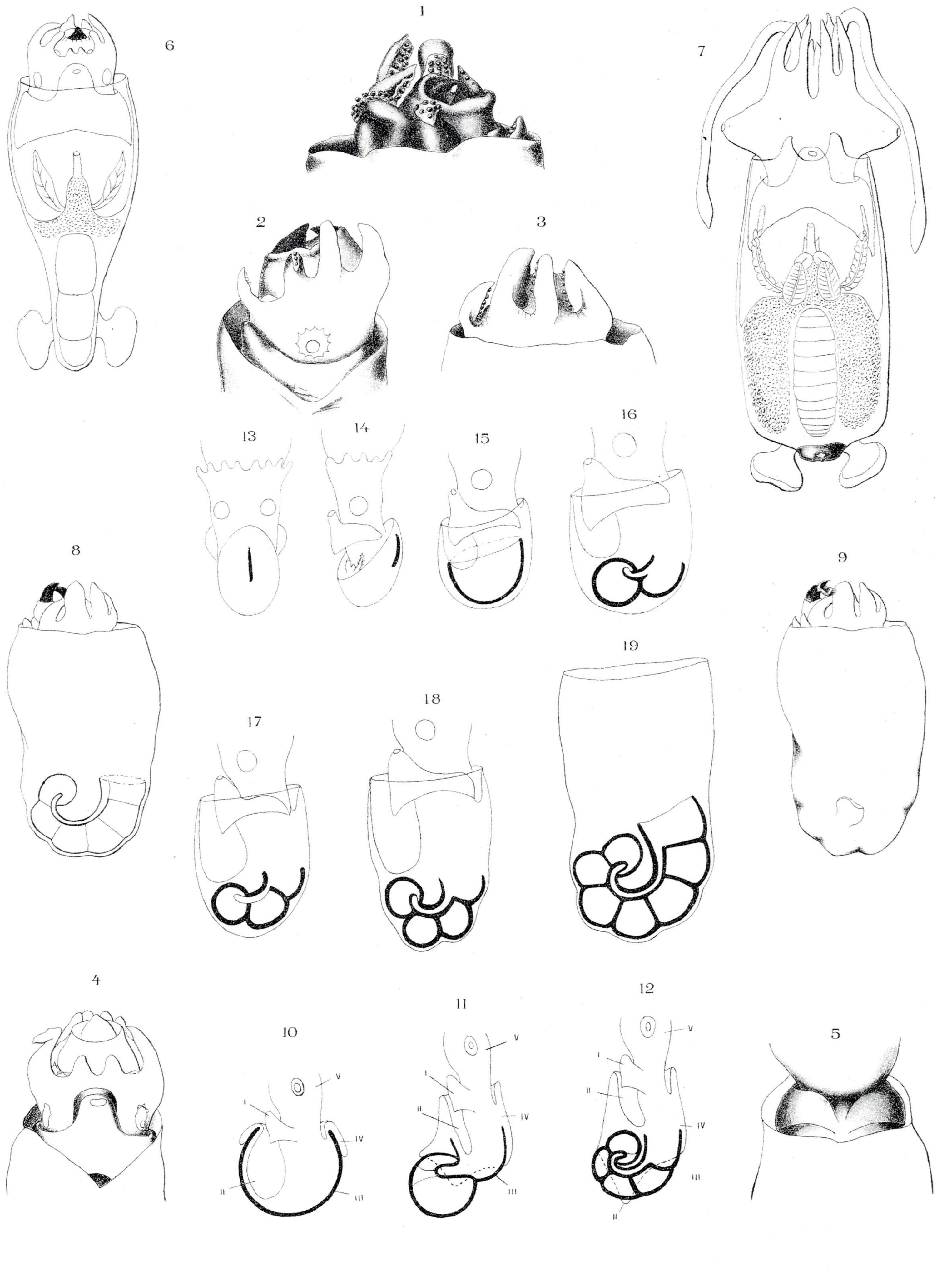

PLATE XVI

Figs. 1–19. *Spirula* sp.

Fig. 1. Details of the part of head projecting from the mantle. × 28.

Fig. 2. View of head after folding down the left margin of the mantle. It also shows the eye, and left half of the funnel.

Fig. 3. Dorsal region of the head showing arms. × 20.

Fig. 4. Ventral view after folding down the mantle margin to show the funnel and eyes. × 18.

Fig. 5. Dorsal view of the funnel after folding down the mantle margin. × 18.

Fig. 6. Reconstituted ventral view of the young *Spirula*. Dotted part represents visceral mass.

Fig. 7. Reconstituted *Spirula* adult. Viscera, particularly the genital organs, are shown dotted. As compared to Fig. 6, the size of the eyes and the volume of the visceral mass are noteworthy.

Fig. 8. Outline sketch of the young *Spirula* in Fig. 9, showing the assumed form of its shell inferred through its transparency.

Fig. 9. Outline sketch of the young *Spirula* based on a photograph, showing where the shell lies on the mantle border.

Figs. 10, 11 and 12. Reproduction of three diagrammatic illustrations by Pelseneer. Fig. 10 represents the author's assumption of the embryonic shell with one chamber. Fig. 11 represents a shell with two chambers. Fig. 12 represents a shell with five chambers.

I*—funnel; II—mantle cavity; II′—terminal disk; III—shell; III′— mantle; V—head.

Figs. 13 to 19. A series of diagrammatic and hypothetical illustrations, explaining the development of the shell. Beginning with Fig. 16, the animal, having grown is less enlarged than in Figs. 13 to 15. The young *Spirula* studied here should be inserted between Figs. 18 and 19.

*Illustrations show I, II, III, IV and V instead of I, II, II′, III, III′ and V.

PART IV

This fascicule was published and sent to the Government of Monaco on April 25, 1924

Results of the Scientific Expeditions

Undertaken on his Yacht
by
ALBERT I
Crown Prince of Monaco

Published under his Supervision
with the help of
JULES RICHARD
Doctor of Science
In-charge of the On-board Zoological Work

FASCICULE LXVII
Contribution to the Study of
Cephalopods of the North Atlantic

Part 4

by *Louis Joubin*

with fourteen plates

Printed in Monaco
1924

3 # Introduction

This volume is the fourth in the series "*Results of the Scientific Expeditions of H. R. H. Prince Albert of Monaco*" devoted to the sudy of cephalopods.* It contains a great deal of new and interesting material on the morphology and biology of these mollusks. The publication of this volume was greatly delayed due to the difficult circumstances we have encountered in recent years.

This beautiful collection of cephalopods, obtained during the cruises of the late Prince Albert, was entrusted to me for study. This study is forcibly incomplete since the species were often represented by single specimens which it was necessary to impact as little as possible so it could be displayed at the Museum of Monaco. In spite of this inadequacy, I think this volume will hold an honorable spot next to the previous volumes.

4 The totality of the cephalopods collected since the commencement of the cruises constitutes a most valuable scientific record on these mollusks. It may be stated that this is the richest collection of cephalopods of the North Altantic that exists. Other collections may have more specimens because they include exotic species. However, I do not think that any other collection contains an equal abundance of forms, known or completely new, belonging to a marine region as well defined as is the North Atlantic. Many of these cephalopods have interesting anatomical features, the study of which has advanced our knowledge of the morphological relations of these animals. Some of the young forms are equally interesting and additional stages in future catches will make them landmarks in a continuous series. With the discovery of more stages, the records, though initially insignificant become more informative. They have the cumulative effect of showing that cephalopods frequently undergo true metamorphosis. Some isolated young collected initially appear to be unique and difficult to affiliate to a species. Such young are later found to be one of the developing stages of an adult long known but whose larval phases were unknown. This is one reason why the synonymy and nomenclature of cephalopods is so confusing. Many names have been given to the same species in its several different stages. It is only by gradually procuring these documents that a link can be established in which the various preliminarily described forms can be intercalated. All this will lead to future clarification and simplification but at present the situation is frequently very complicated.

*The preceding three are: fasc. IX, 1895; fasc. XVII, 1900; fasc. LIV, 1920.

5

Classification

Cephalopoda

SUBORDER OCTOPODA LEACH, 1818

Division A. LIOGLOSSA Lütken, 1882

Family I. PTEROTI Reinhardt and Prosch, 1846

Genus *Cirroteuthis* Eschricht, 1836

1. *Cirroteuthis umbellata* P. Fischer, 1883

Division B. TRACHYGLOSSA Lütken, 1882

Family II. OCTOPODIDAE d'Orbigny, 1838

Genus *Octopus* Lamarck, 1799

2. *Octopus vulgaris* Lamarck, 1799

Family III. ELEDONIDAE Joubin, 1918

Subfamily Eueledonidae Joubin, 1918

Genus *Graneledone* Joubin, 1918

3. *Graneledone verrucosa* (Verrill), 1918

Subfamily Bolitaenidae Chun, 1906

Genus *Vitreledonella* Joubin, 1918

4. *Vitreledonella richardi* Joubin, 1918
5. *Vitreledonella alberti* Joubin, 1919

Genus *Bolitaena* Steenstrup, 1859

6. *Bolitaena diaphana* (Hoyle), 1885

SUBORDER DECAPODA LEACH, 1918

Division A. MYOPSIDA d'Orbigny, 1845

Family I. SEPIDINI Steenstrup, 1801

Genus *Heteroteuthis Gray, 1849**

7. *Heteroteuthis dispar* (Rüppell) Gray, 1849

6 **Genus *Rossia* Owen, 1834**

Subgenus *Franklinia* Norman, 1890

8. *Rossia caroli* Joubin, 1902
9. *Rossia palpebrosa* Owen, 1835

Division B. OEGOPSIDA d'Orbigny, 1845

Family I. ONYCHOTEUTHIDAE G. Pfeffer, 1900

Genus *Onychoteuthis* Lichtenstein, 1817

10. *Onychoteuthis banksi* (Leach) Férussac, 1817

Family II. ENOPLOTEUTHIDAE G. Pfeffer, 1900

Genus *Abraliopsis* Joubin, 1896

11. *Abraliopsis pfefferi* Joubin, 1896
12. *Abraliopsis* sp. juv.

Genus *Abralia* Gray, 1849

13. *Abralia morrisii* Vérany, 1837

Genus *Thelidioteuthis* G. Pfeffer, 1900

14. *Thelidioteuthis alessandrinii* (Vérany), 1851

Genus *Pyroteuthis* Hoyle, 1884

15. *Pyroteuthis margaritifera* (Rüppell), 1844
16. *Pyroteuthis margaritifera,* var. *aurantiaca* Joubin, 1919
17. *Pyroteuthis* juv.

*Incorrectly identified in original text as *Heroteuthis*—Scientific Editor.

Family III. PARATEUTHIDAE, Joubin 1919

Genus *Cycloteuthis* Joubin, 1919

18. *Cycloteuthis sirventi* Joubin, 1919

Family IV. HISTIOTEUTHIDAE Verrill, 1881

Genus *Meleagroteuthis* G. Pfeffer, 1900

19. *Meleagroteuthis hoylei* G. Pfeffer, 1900

7 **Genus *Calliteuthis* Verrill, 1880**

20. *Calliteuthis meneghinii* (Vérany), 1851
21. *Calliteuthis* sp. juv.

Subgenus *Stigmatoteuthis* G. Pfeffer, 1900

22. *Stigmatoteuthis verrilli* G. Pfeffer, 1912

Genus *Histioteuthis* d'Orbigny, 1839

23. *Histioteuthis bonelliana* Férussac, 1835

Family V. GONATIDAE Hoyle, 1886

Genus *Gonatus* Gray, 1849

24. *Gonatus fabricii* (Lichtenstein), 1818

Family VI. BENTHOTEUTHIDAE G. Pfeffer, 1912

Genus *Benthoteuthis** G. Pfeffer, 1912

25. *Benthoteuthis megalops* Verrill, 1895

Family VII. BRACHIOTEUTHIDAE G. Pfeffer, 1912

Genus *Tracheloteuthis* Steenstrup, 1881

26. *Tracheloteuthis riisei* (Steenstrup), 1852

Family VIII. OMMATOSTREPHIDAE Steenstrup, 1861

Genus *Ommatostrephes* d'Orbigny, 1835

27. *Ommatostrephes sagittatus* Lamarck, 1799

*Incorrectly identified as Benthoteuthidae in text—Scientific Editor.

Genus *Illex* Steenstrup, 1880

28. *Illex illecebrosus* (Lesueur), 1821

Genus *Stenoteuthis* Verrill, 1880

29. *Stenoteuthis bartrami* (Lesueur), 1821

30. *Stenoteuthis pteropus* Steenstrup, 1855

Genus *Rhynchoteuthion* G. Pfeffer, 1900

31. *Rhynchoteuthion* sp. juv.

Family IX. CHIROTEUTHIDAE Gray, 1849

Subfamily Chiroteuthinae G. Pfeffer, 1912

Genus *Chiroteuthis* d'Orbigny

32. *Chiroteuthis portieri* Joubin, 1912
33. *Chiroteuthis veranyi* (Férussac), 1835
34. *Chiroteuthis picteti* Joubin, 1894
35. *Chiroteuthis lacertosa* Verrill, 1881

Subfamily Mastigoteuthinae G. Pfeffer, 1912

Genus *Mastigoteuthis* Verrill, 1881

36. *Mastigoteuthis grimaldii* Joubin, 1895
37. *Mastigoteuthis cordiformis* Chun, 1908

Family X. CRANCIIIDAE G. Pfeffer, 1900

Genus *Megalocranchia* G. Pfeffer, 1884

38. *Megalocranchia hyperborea* (Steenstrup), 1856
39. *Megalocranchia abyssicola* Goodrich, 1896

Genus *Corynomma* Chun, 1906

40. *Corynomma speculator* Chun, 1906

Genus *Galiteuthis* Joubin, 1898

41. *Galiteuthis armata* Joubin, 1898

9 # Descriptive Part

Cephalopoda

SUBORDER **OCTOPODA** LEACH, 1818

Division A. LIOGLOSSA Lütken, 1882

Family I. PTEROTI Reinhardt and Prosch, 1846

Genus *Cirroteuthis* Eschricht, 1836

Cirroteuthis umbellata P. Fischer, 1883

1883. *Cirroteuthis umbellata* P. Fischer (**19**), p. 402
1900. *Cirroteuthis umbellata*, L. Joubin, (**35**), p. 21, Pl. 1
1904. *Stauroteuthis hippocrepium* Hoyle (**27**), p. 6, Pl. 2 and 3

Expedition of 1913: Stn. 3412, depth 5,270 m. Augut 9, 39°59′ N., 56°20′ W. Trawl. Sigsbee Trench south of Newfoundland.

The specimen is in good condition. It is a young individual approximately 10 cm long. The fins, inserted considerably anteriorly, are approximately 23 mm long from the point of their insertion to their tip. The interior of the web is brick-red. The outside of the body is grayish with red spots, plus red on the web. The entire body is semitransparent, gelatinous and very soft.

Division B. TRACHYGLOSSA Lütken, 1882

Family II. OCTOPODIDAE d'Orbigny, 1838

Genus *Octopus* Lamarck, 1799

Recent authors, following Hoyle, have changed the old name of *Octopus* to *Polypus* on the basis of priority as per the rules of nomenclature. I have retained the name *Octopus* to denote the common animal of our coasts and all its congeneric forms since I am not convinced of the importance of the reasons given. I see no advantage in changing a name accepted by all naturalists for more than a century or to replace it
10 by another under the pretext that an older name has been found in an old paper although it is totally unknown. I am quite convinced that the name *Polypus* will cause confusion because it has been used to denote coelenterates.

Octopus vulgaris Lamarck, 1799

Expedition of 1911: Stn. 3122. August 12. Southern tip of Desert Grande Island. Caught at low tide. One average-size male specimen.

This is a common species and inhabits the entire littoral area of the European coast to Scotland, including the Mediterranean and the west coast of Africa, the Azores and the Canaries.

Family III. ELEDONIDAE Joubin, 1918

Subfamily Eueledonidae Joubin, 1918

Genus *Graneledone* Joubin, 1918

Graneledone verrucosa (Verrill), 1881
(Pl. IX, Figs. 1–3; Pl. VI, Figs. 15–17)

1881. *Eledone verrucosa* Verrill (**82**), p. 105
1881. *Eledone verrucosa*, Verrill (**83**), p. 380
1882. *Eledone verrucosa*, Verrill (**84**), p. 183
1884. *Eledone verrucosa*, Verrill (**85**), p. 148
1917. *Moschites verrucosa* (Berry) (**5**), p. 1
1918. *Moschites verrucosa* var. *media* (Joubin) (**43**), p. 5
1918. *Graneledone verrucosa*, Joubin (**44**), p. 39

Expedition of 1913: Stn. 3437, depth 1,458 m. 42°40′ N. 62°49′30″ W. Trawl. Southwest of Halifax. A single complete specimen in perfect condition.

The description of this very interesting cephalopod must be prefaced by a brief historical review of the variable versions of this species, of which only a few specimens are so far known.

One male and one female specimen were caught during the voyages of the *Blake*, under the command of Professor A. Agassiz, in two localities near the *Prince of Monaco* station 3437 mentioned above. One was caught at 810 fathoms and the other at 466 fathoms (1,483 and 853 m).

The total length of the male was 202 mm and the female was 360 mm.

Professor Verrill gave a brief description (**82**) of these animals accompanied by two plates showing the dorsal side, profile and hectocotylus of the male. They are too schematic.

The same memoir incorporating the same plates (**83**) was merely reproduced by Professor Verrill in the
11 Transactions of the Academy of Sciences of Connecticut in 1881 and in 1882 in the Annual Report of the Commissioner of Fish and Fisheries for 1879.

In his Second Catalogue of Mollusca (**85**) in 1884, Verrill reported several other specimens: a large male, 292 mm long, caught aboard the *Fish Hawk* in 1882 in the open waters off Nantucket at 787 fathoms (1,442 m). In 1883, during the cruises of the *Albatross*, four more specimens were caught at 1,050, 1,106, 1,255 and 1,209 fathoms (1,921, 2,024, 2,297 and 2,213 m), probably near the preceding stations. The author gives a brief description of a large male but does not mention the others except that three of them have the same features as the large male but are smaller and have less protuberant warts.

In 1886, Hoyle (**24**), in his beautiful memoir on the cephalopods of the *Challenger*, assigned the name *Eledone verrucosa* Verrill to a specimen caught in the open waters of the Kermadec Islands. Thereafter, as per the code of Zoological Nomenclature, the old name of *Eledone*, universally known to all naturalists, has been replaced by *Moschites*.

In 1916*, S. Berry (**5**) expressed surprise at the occurrence of the species around the Kermadec Islands in the open Pacific. Up to that time it had been found exclusively on the east coast of the United States. Therefore,

*Cited as 1917 in original (see Bibliography)—Scientific Editor.

he restudied this cephalopod. He decided that Hoyle's step of synonymizing his *Eledone verrucosa* with Verrill's was not proper. In a recent note (7) published in January 1917, he spelled out his reasons and stated the necessity for maintaining the name *Moschites verrucosa* for the Atlantic species described by Verrill in 1881, and for assigning the name *Moschites challengeri* to the Pacific species of the Kermadec Islands described by Hoyle in 1886.

It is also noted that Hoyle (27) reported three specimens caught in the Gulf of Panama at 1,020 fathoms (1,866 m) during one of the cruises of the *Albatross* in the Pacific. The poor condition of the material did not permit description and illustrations; however, these specimens were probably not adults. Mr. Berry suggested that they might be a different species which, in any case, should not be affiliated to *M. verrucosa* of Verrill 1881.

The original illustrations of *M. verrucosa* Verrill are confined to dorsal and lateral views given by the author and reproduced in his various memoirs since 1881, as well as the too diagrammatic representation of the hectocotylus. In 1917, Mr. Berry gave two photographs of the hectocotylus of the type specimen of *M. verrucosa* Verrill in the Museum of Comparative Zoology and of the same organ in *M. challengeri* Berry (ex *verrucosa* Hoyle) in the British Museum. One could thus see the difference between the two organs. Moreover, Mr. Berry had photographed both Verrill's and Hoyle's original specimens. The two photographs
12 are of the dorsal view of the animals to show the distribution of the papillae on the dorsal skin. The two figures are very similar, the only difference is that the papillae in *M. verrucosa* Verrill are more widely spaced and do not extend anteriorly toward the mantle and head, whereas in *M. challengeri* they are set more closely together and extend onto the bases and web of the dorsal arms. The actual difference between the two species therefore lies in the hectocotylized arm; a large web in *M. challengeri* and no web in *M. verrucosa*, and the former found in the Pacific and the latter in the Atlantic.

It seemed the problem was resolved by Mr. Berry's memoir. However, the study of the specimen caught in perfect condition by the Prince of Monaco during its cruise of 1913 has reopened the issue.

First, this specimen is unique and there can be no possibility of confusing it with another. It is a highly developed completely intact adult male. Its color is natural and there is a little distortion in the mantle which was posteriorly flattened by its storage at the bottom of the jar. This specimen was caught in a trawl from the 1,458 m sea bottom southwest of Halifax (42°40′ N, 49′30″ W), approximately in the latitude of Cape Cod. This is undoubtedly an Atlantic species found relatively far removed from the areas where the specimens described by Verrill were caught.

An unexpected point arises here: the hectocotylus resembles but is not identical to Berry's depiction of the Pacific species caught around the Kermadec Islands and named *M. challengeri.* The distribution of papillae on the dorsum distinctly resembles that of *M. verrucosa.* The only difference is that the papillae progressively diminish in number and size as they approach the base of the arms, the last ones just reaching the beginning of these arms. Apart from this detail, the photograph of the *M. verrucosa* given by Mr. Berry unquestionably resembles the specimen in the collection of the Prince of Monaco.

The specimen can be affiliated to *M. verrucosa* Verrill on the basis of its papillae, while its hectocotylus affiliates it to *M. challengeri* Berry.

The only two logical conclusions that can be drawn are:

1. To erect a new species that will be an intermediate between the two mentioned above.

2. To consider that these papilliferous eledonids are geographic varieties of one species, of which we now know three. I cannot elaborate further for lack of records from various localities that might definitely solve the problem. Moreover, the study made by Mr. Berry of the hectocotylus of *M. verrucosa* is probably inadequate. The end is contracted and it is possible that the palm [ligula] has been modified by its prolonged preservation in alcohol.

In conclusion, I feel I must establish the following list of varieties:

13 1. *Moschites verrucosa* Verrill, original type, Atlantic.
2. *Moschites verrucosa* var. *challengeri* Berry, Pacific.
3. *Moschites verrucosa* var. *media* Joubin, Atlantic, intermediate between the two others.

Furthermore, the study of the other species of closely related cephalopods has resulted in the separation of the genera to which they belong and in the establishment of the genus, *Graneledone*, to which I affiliate the present species [*G. verrucosa* (Joubin) 1918].

Details of the specimen examined (Pl. IX, Fig. 1) are given below.

Body heavy, compact, strewn with papillae on the dorsum. Basally thick arms form a solid crown, broader than the head and body, with tips curled; no constriction to indicate neck between the head and body.

Overall grayish-violet on the dorsal side, reddish-violet gray on the ventral; the arms slightly lighter colored at their tips due to an enormous number of very small, closely packed, chromatophores.

The maximum diameter of the base of the arms is 110 mm; the constricted part between the arms and the eyes is 85 mm; width of the head at the level of the eyes is 95 mm; maximum width of the mantle, almost at the middle is 105 mm. Length of the mantle, below the brachiocephalic constriction is 102 mm.

Length of the arms (mm):

1st arm (dorsal) right	262
2nd arm	255
3rd arm	265
4th arm (ventral)	270
1st arm (dorsal) left	270
2nd arm	280
3rd arm	280
4th arm (ventral)	270

It is observed that the left arms are longer than those on the right but the differences between all these arms are too small to draw conclusions on their ratios of length.

The rather deep web uniting the bases of these arms can be measured from the brachiocephalic furrow delimiting the head above the eyes. It is rather difficult to give exact measurements but it is visible enough to measure the distance separating the furrow from the base of the depression between two juxtaposed arms. This distance is 65 mm on the dorsomedian line, 80 mm between arms 1 and 2, 95 mm between arms 2 and 3, 105 mm between arms 3 and 4 and 100 mm on the ventromedian line.

Each arm has 60 large, medium and small suckers with approximately 20 minute suckers at the tip. This number varies only on the hectocotylized arm, where there are 42 large and medium suckers and no small suckers.

14 One very peculiar distinguishing feature is found in this cephalopod. It is known that sometimes, but very rarely, a bifurcated arm is found in some cephalopods. A few such cases have been reported. In this specimen, the dorsal right arm is bifurcated at the 47th sucker; from here, each of the two almost equal branches is from 22 to 24 mm long and has 30 suckers, ten of which are very small (Pl. IX, Fig. 1).

The bases of the arms, the top of the head and the mantle form an almost flat surface with a slight median depression due to the extension of the groove separating the two dorsal arms. The commissures of the mantle opening are entirely ventral and cannot be seen from the back. The body is posteriorly globular. This distinguishing feature is not clearly visible in the illustrations given by Verrill or the photographs of the Verrill and Hoyle type specimens given by Berry. This is probably because these cephalopods were subjected to long storage in alcohol which distorted one of the most important characteristics found in the specimen in the collection of the Prince of Monaco. That specimen was preserved in formalin. The tissues of these animals are permeated with water puffy and edematous which has rounded the body, head, arms and the margins of the mantle. Handling the specimen to measure or photograph it squeezed this water out of the skin and its resemblance to the slender specimens preserved in museums could be seen. This observation was made because the initially puffy and smooth skin became wrinkled and the arms seemed to be thinner. The very strong musculature of the arms and body could thus be seen and is certainly more marked than in the *challengeri* photographed by Mr. Berry.

The dorsal side of the head, body and bases of the dorsal arms are strewn with papillae formed by the union of small juxtaposed warts. The largest ones are between the eyes, on the top of the head and mantle. They decrease with distance from these regions, are very small on the border of the head, mantle and beginning of the dorsal arms, and soon totally disappear. Thus, there are no papillae on the ventral side of the animal or on the free part of the arms; they are present only at the bases of the four dorsal arms.

The largest papillae form a circle around the eyes not far from the encircling, thick, deep-red eyelid. The two largest papillae are dorsally located on the top of each eye, the succeeding papillae on the sides are smaller, and those below the eye are very small. There are a few minute papillae below the larger papillae within this circle. The palpebral circle contains 12 papillae. Ventrally two other incomplete circles surround it. All these large or small dorsal papillae are very regularly spaced 6 to 8 mm apart. Rudimentary ones are occasionally found interspersed among the others.

The farthest, most marginal papillae at the edge of the head, base of the arms, and dorso-ventral limit of the mantle are only tiny cutaneous elevations. Further inward they combine and form three or four small warts; the largest papillae consist of four to eight of these warts. Around the eyes there are two large papillae-like horns of around 20 warts; one at the top is white and more developed than the others.

The ventral side is smooth, puffy and soft. The edge of the mantle opening is not thin and distinct like in other cephalopods but so swollen and soft that this large oedematous cushion is impossible to accurately delineate. The funnel is small, its tip is yellowish and the opening is square; the chromatophores do not reach to the opening. The lateral pouches of the funnel are poorly developed.

Now some observations on the hectocotylus on the third right arm. It is highly developed and, as has been noted, more closely resembles that of *Octopus* than that of *Moschites*. The length of the terminal hectocotylized part is 13 mm from the extreme tip to the middle of the sucker immediately preceding it. This spatulate, heart-shaped, section is 10.5 mm long and 7 mm at its greatest width. This large section, ordinarily lined with transverse furrows, is completely smooth here and the lighting has to be varied to be able to distinguish some lines so faint that they are almost impossible to count. The surface is slightly concave, with slightly protuberant pads at the edges; a few isolated chromatophores are distributed on the edge. A well-developed spur hollowed into a groove rises from the base of the spoon. It has two white diverging edges which define the V-shaped form of the groove. One branch of the V loses itself in the skin to the left of the first sucker; the right branch continues with a long groove bordered by a strong cutaneous fold along the entire length of the arm to the web between the third and fourth right ventral arms. The lip defining this groove is very marked and covers a strong gutter particularly toward the base.

As can be seen in the figures (Pl. VI, Figs. 16 and 17), the hectocotylized arm is bent by the strong web that is much better developed than in the other arms, and of which the spermiferous groove forms the free edge.

Genus *Vitreledonella* Joubin, 1918

Vitreledonella richardi Joubin, 1918
(Pl. I, Figs. 1–2; Pl. III, Fig. 1; Pl. IV, Figs. 1–9; Pl. V, Figs. 1–20; Pl. VI, Figs. 1–14)

1918. *Vitreledonella richardi* Joubin (**44**)

Expedition of 1912: Stn. 3223, depth 5,300 m. August 10. 30°50′ N, 25°43′ W. Trawl. Monaco Trench.

16 This cephalopod is one of the most interesting that has been collected during the cruises of the Prince of Monaco.

In designating it I have dedicated this strange octopod to my friend Dr. J. Richard, Director of the Museum of Oceanography of Monaco.

The sole specimen was unfortunately so badly damaged while trawling that it has not been possible to make a complete study. Its body is extremely soft and the other objects caught in the net had mutilated it.

Only five arms are still attached to the head; two others, still united, are completely separate from the rest of the body. A small fragment of the eighth arm was found detached from all the rest. The dorsal part of the brachial crown is entirely open as the three arms that had been ripped off had left the buccal bulb exposed. The funnel is also split into two parts and the head is only held to the mantle by a shred of skin.

It is necessary to mention these defects to explain the lacunae in the description. However, on joining the parts still in place and identifying, as far as possible, the detached pieces, I was able to reconstruct the animal, albeit incompletely. Due to the talent of Miss Vesque the illustration of this cephalopod closely resembles its natural state (Pl. I, Figs. 1 and 2).

It is observed that the animal has only seven arms. I was not sufficiently certain of the position of the 8th to show it in the figure. Its base was completely torn away down to the buccal bulb and the only remaining evidence of it is a fragment of its middle. I do not have accurate proof of its position, although, by process of elimination, its position is almost certain as will be indicated later in the text.

The fact of finding a piece of this eighth arm has some importance for one might think that this animal normally does not have a hectocotylized arm. This hypothesis which furthermore is not likely for an eledonid is not affected by the presence of this fragment; moreover, the remnants of the viscera have shown that the specimen is a female.

An interesting fact to be observed is that this cephalopod was caught in a trawl in 5,300 m of water, i.e., on the seabed (Stn. 3223. Monaco Trench, south of San Miguel Island, Azores). Its extreme transparency and great fragility might lead one to think that it is a pelagic animal. It could have been caught while the net was being hauled up, but, in view of the mode of operation of this gear, this is unlikely. It is observed that its musculature is highly reduced, its arms are not linked by a very large, strong swimming membrane. Although its funnel is rather large it is not firm, the mantle is wide open, thick and soft; all these features do not denote a vigorous swimmer.

17 Under these conditions, it might be considered to be an animal that roves on the seabed, but susceptible to rising to search for pelagic prey. It cannot feed on large animals because its softness would prevent it from chasing energetic prey. The weakness of its brachial muscles and its suckers and the reduction of its buccal apparatus make it almost unarmed. It can only feed on small plankton or soft matter disintegrating in the mud.

Exterior: The entire animal is gelatinous, almost translucid, particularly the arms which are completely hyaline. The head and body are less transparent because they are thicker, but the viscera are still visible through the skin and muscles. The chromatophores, very rare on the arms except on the suckers, are more abundant on the dorsal side of the head and mantle. The overall color of the skin of the specimen preserved in formalin is pale yellow.

The arms, described later in the text, are long, cylindrical, transparent, differ only slightly from each other in size and shape, and have a very weak musculature. They have a single row of small, widely spaced suckers at their bases which become closer and larger in the middle and very small and contiguous at the tip of the arms.

The brachial crown rests on a large, transversely developed head with very weak vertical development, so that the head is six times broader than its length. It is transparent and the brain and nerves issuing from it are visible. The eyes are large, black, laterally flattened and semicircular when seen from the back. Their anterior side is rectangular when seen in profile, with a large lens and an exteriorly open corneal duct.

The visceral mass or body is posteriorly rounded, bursiform, almost as broad as long without a trace of fins. The mantle opening seems to be well-developed at least as broad as the body, but its lateral limits are difficult to define because of the poor condition of the specimen.

The funnel is of medium size; its two lateral pouches are well-developed.

The web can be seen between the arms. It is extremely delicate, transparent and soft, but very difficult to discover where it ends as it is intact only between the two ventral arms and torn everywhere else.

This animal resembles in shape, size and proportions the common elédonid of the Mediterranean, but differs in its transparency.

Arms: The measurements given below can only be considered approximate because it is difficult to measure the incomplete, torn, and severed arms. From the above it can be assumed that these arms do not differ very greatly from each other. They are so soft that the mere act of laying them on a graduated ruler to measure
18 them can cause a variation of several millimeters.

1st arm (dorsal) right	205 mm
2nd arm	missing
3rd arm	182 mm
4th arm (ventral)	truncated
1st arm (dorsal) left	204 mm
2nd arm	156 mm
3rd arm	175 mm
4th arm (ventral)	158 mm

The brachial cylinder, in circular section, consists of a fibrous tube with parallel longitudinal muscular fibers arranged in an extremely thin layer. This musculature is extremely weak.

The transparent smooth skin covering the arms consists of connective tissue with an irregular network. It is very loose, permeated with water and has few rare chromatophores.

The fibrous cylinder is filled with transparent connective tissue, with slender trabecular fibers. The brachial nerve runs along the center of this cylinder regularly following its entire longitudinal axis. The nerve enlarges into a ganglion below the level of each sucker. Its description is given later in the text.

The web membrane is well marked between the arms but has been torn everywhere except between the two ventral arms (Pl. IV, Fig. 1) where it is whole. A deep depression can be seen here which, judging from the traces of it between the other arms, must have been deeper in the former than in the latter. Between the two ventral arms the base of the depression is 35 mm above the joint at the base of the fourth arm. Between ventral-lateral arms four and three, the depression seems to be 65 mm above the corresponding point; this seems to be true for the others. The web between the dorsal arms in Fig. 1 is deduced from indications and fragments but its shape is indistinct.

The web between the two ventral arms (Pl. IV, Fig. 1) is interesting. It consists of a cutaneous layer formed by the accretion of extra skin and the intra-umbrellar skin; but these two layers are absolutely transparent. The cutaneous muscular fibers, which must also have been transparent when alive, have become opaque and white due to the action of formalin making their distribution very distinct. One sees here a first group parallel to the free edge of the membrane but losing itself higher up on the two neighboring arms; lower down some transverse fibers proceed from one arm to the other and become decreasingly noticeable as they approach the peribuccal base of the membrane. Other fibers perpendicular to the others link the free edge of the web with the base of the interbrachial angle. This type of structure makes the web very slightly resistant and it can scarcely be considered to be a swimming organ; it is more likely a funnel apparatus, aiding in the capture of planktonic organisms.

19 *Suckers*: I counted 30 suckers on the intact dorsal arm. They are strictly arranged in a single straight line and at no point deviate in a zigzag, as is frequently seen in eledonids.

They are widely spaced and small (Pl. IV, Fig. 1) in the proximal part of the arm starting from the mouth. They then become larger and more closely spaced from the 9th sucker, almost in the middle of the arm. The three largest are the 12th, 13th and 14th suckers, in approximately the third quarter of the arm. Thereafter, they become increasingly closer and the last few suckers are tiny, touching each other all the way to the tip of the arm.

The first sucker of each arm is slightly larger than the following: together they form a circlet of eight suckers around the peribuccal lip, slightly more protuberant than the suckers that follow.

The margin of each sucker was possibly circular in the live animal, but due to the action of formalin this margin has been folded into four lobes leaving a gap between them at four places (Pl. IV, Fig. 2).

Each sucker is implanted on a transparent, hemispherical nipple where its opening occupies the peak and descends almost to two-thirds of its height (Pl. VI, Fig. 14).

The sucker itself is slightly less transparent than the nipple holding it, perhaps due to the action of the preservative which clearly brings out the details of its surface. It is shaped like a small cup with a broad edge and pointed base, a constriction and a slight groove almost in the middle of the organ below the broadened edge (Pl. IV, Fig. 3). A circular opening corresponds to the groove and leads into an oval-shaped cavity forming the base of the sucker. This cavity contains no trace of an organ resembling a piston as found in the suckers of other cephalopods. The wall is smooth, regularly oval-shaped and has no folds or protuberances on the conical base in the cavity of the cup.

The body of the sucker is light yellowish and its outer surface is strewn with yellow and brown chromatophores of different sizes, some spread out and others contracted.

The suckers closest to the mouth are 2 to 2.5 mm in diameter; those in the middle, the largest, are approximately 4 mm; they are probably slightly larger in the live animal when the edges are not folded inward. The distalmost suckers at the tip of the arms are scarcely more than 1 mm in diameter.

The nipple holding the sucker has a rather peculiar structure (Pl. IV, Fig. 3, and Pl. VI, Fig. 14): it consists of a soft, transparent protuberance of the skin, but is supported by a sort of fibrous conical body issuing from the brachial cylinder. The cup of the sucker is positioned inside the upper opening of this conical body whose broad base is in direct contact with the arm's cylinder of muscular fibers. The interior of the support is filled with gelatinous hyaline tissue in which nerves pass to the sucker from the corresponding ganglion.

20 The innervation of the suckers is very remarkable. The brachial nerve (Pl. VI, Fig. 14) runs along the length of the short arm and is flanked by two vessels, probably one vein and one artery. A large oval-shaped ganglion (Pl. IV, Fig. 3*, *G* and Pl. VI, Fig. 14) traversed by the brachial nerve is observed slightly below the level of each sucker. This nerve successively traverses all the ganglia of the arm. Near the peak of the ganglion a nerve branches out to proceed on a winding trajectory toward the base of the sucker and reaches it after crossing a very small ganglion in close proximity to it (Pl. IV, Fig. 3 and Pl. VI, Fig. 14, P).

Other nerves also issue from the ganglion, some toward the muscular cylinder, others penetrating the support of the sucker; a good number of the latter nerves reach the groove of this organ and appear to penetrate it. However, some of them seem to reach the chromatophores. It is extremely difficult to precisely define these features in view of the failure to fix the fresh tissues. However, it can be affirmed that the largest part of the brachial nerve aids the functioning of the corresponding sucker.

Sections made of the brachial ganglion provide the following information (Fig. 2, Pl. V). A cortical cellular layer *C* and a central fibrous mass *F* can be distinguished. The section is rounded at the top and flattened at the base; the main nerves branch out at the angles of this face. The origin of the fibers constituting them can be followed quite far into the central substance.

*'*G*' not indicated in this figure—General Editor.

The cortical layer is formed by a large number of small cells reacting strongly to coloration by hematoxylin. The nerve cells *G* are much larger in the upper region below the artery *A*. They cover a portion of the differentiated fibrous mass *T* corresponding to Guerin's myelinic zone which he has reported in octopods and decapods.

Nuclei, probably from the intra-nerve connective tissue, are disseminated in the fibrous mass.

The entire ganglion is enveloped by a connective tissue layer which also encloses the artery *A* and the longitudinal vein *V*.

The section of the axial nerve (Fig. 3, Pl. V) between the ganglia shows the presence of nerve fibers forming an irregular matrix with clear spaces. Here and there are connective tissue nuclei and finally the artery *A* and the brachial vein *V*.

Histological structure of the sucker: As mentioned earlier in the text, the sucker has the shape of a deep cup with flaring edges. The wall is almost of the same thickness throughout; it is observed, however, that it is slightly thicker in the deeper part and in the rim region. The main mass of the sucker consists of muscular tissue (Pl. V, Fig. 4) arranged in bundles of perpendicular fibers *P* from the outer to the inner wall. These fibers are intermingled with other less numerous circular fibers which are more abundant in the lip. Connective
21 tissue is also found here. The inner surface of this trabecular layer is covered by a layer of muscles separated by oblique, intercrossed and seemingly braided bundles *O*. A similar but very much thinner layer *M* covers the outside of the sucker. Thus there are three muscular layers: a thin outer layer of oblique and circular fibers, the middle layer which is thickest and consists of trabeculae mixed with circular fibers and an internal layer of oblique bundles.

The circular muscle fibers are more abundant in the labial region (Pl. V, Fig. 5) and the wavy surface on the outer edge of the lip has numerous longitudinal fibers in bundles *L*.

The entire internal surface of the sucker is lined with a flat, continuous epithelium *E* with small uniform cells containing a large nucleus.

Nowhere is there a trace of a muscular swelling which constitutes the piston found in other cephalopods. On the contrary, the base seems to be thinner than the middle region and definitely does not function in this manner.

The head is remarkable for its great width as compared to its length, at a ratio of 6:1. This structure results in a considerable separation of the eyes and the resulting lengthening of the optic nerves. On removing the skin, it can be seen that the subjacent tissues are absolutely transparent which distinctly expose the nerves and central nervous system without the necessity of dissection. These tissues have become opaque due to the action of the liquid preservative. The head contains no solid cartilage; there is at the most only a certain quantity of fibrous tissue around the brain, forming a kind of a cage partially enclosing the brain but without the consistency of the cranium of other cephalopods. No distinct, more consistent form supporting the entire head can be defined. The eyes are embedded in this hyaline tissue at both ends of a cephalic cylinder. The arms do not appear to be inserted by their bases into the head, but the poor condition of the specimen prevented me from detecting the exact structure of this region. Three or four very small pink masses are in the hyaline tissue on each side of the brain; these are probably of the same nature as the "white bodies", occasionally pink, found behind the eyes.

Central nervous system: The cerebral mass in the center of the head is suspended by nerves issuing from it into a fibrous capsule (Pl. V, Fig. 6, *F*) which will be discussed later. It is bathed in a semiliquid, gelatinous substance which forms a sort of cushion for it.

A marked distinguishing feature of this nervous system is its smallness in relation to the size of the animal. There is only 3.5 mm between the two emergences of the optic nerves and 5 mm at its greatest vertical dimension. These measurements must be tripled for a coastal eledonid of the same size.

22 The great length of the optic nerve is absolutely characteristic of this form and definitely bars its affiliation to *Bolitaena diaphana*. On the contrary, this feature likens it to *Eledonella pygmaea* whose nerve is almost

as long. However, there is a basic difference between the latter species and this form of *Vitreledonella*. A peduncular ganglion topped by a small oval-shaped spot in *Eledonella pygmaea* is almost in the middle of the optic nerve. In *Vitreledonella*, this peduncular ganglion (Pl. V, Fig. 6, *W*) is not found on the optic nerve, but further away and, with its oval spot, is applied against the optic ganglion *G* where it penetrates the optic nerve.

The entire brain of *Vitreledonella* resembles more closely that of Chun's *Eledonella* than of *Bolitaena*. However, in the latter genus, the small peduncular ganglion is fixed against the optic ganglion as in *Vitreledonella*. In the latter form, there is thus a mixture of the characteristics of the other two genera. One might say that the brain of *Vitreledonella* is analogous to that of *Eledonella* except that the peduncular ganglion has shifted along the optic nerve to lie against the optic ganglion; this is comparable to the case of *Bolitaena* where the optic nerve has been considerably extended.

It has been necessary to stress these relations because they constitute important characteristics to differentiate the three genera, independent of other observed structural differences.

The surface of the brain is covered with a layer of yellow, brown and red chromatophores lying directly on the ganglia. They are very difficult to remove in order to distinguish, by dissection, the points of issue of the nerves. They also mask the details of the cortical layer of the brain.

The three ganglia, the cerebral, pedal and visceral, are rather easily differentiated, as is the buccal ganglion situated immediately ahead of the cerebral ganglion, but it is very difficult to distinguish the nomenclature of these nerves since most of them are broken and cannot be traced to their ends. However, I have been able to identify the main nerves; the others have been named only by analogy with those described by Chun for the other two genera.

The entire brain can be compared to a fragment of a tube (Pl. IV, Figs. 4 and 6) with thick walls and narrow opening bevelled in front of and behind the cerebral ganglion *C*. The latter is dorsal and much smaller than the ventral part (Pl. VI, Figs. 8 and 9). The latter shows the cornet-shaped pedal ganglion *P* and behind it the triangular visceral ganglion *V**. The esophagus *O*, the salivary duct *S* and vessels (two veins and an artery) pass through the tube.

Three nerves of arms 4, 3 and 2 issue from the anterior margin of the pedal ganglion. The nerve of arm 1 has its origin higher and closer to the insertion of the buccal ganglion *B*.

The cerebral ganglion (Pl. IV, Fig. 4 and Pl. VI, Fig. 9) consists of two main masses, one anterior in the form of a transverse pad and a trilobed posterior part. The two large symmetrical optic nerves *R* issue from this ganglion.

23 The buccal ganglion (Pl. IV, Fig. 4, *B*) is in a bilobed pad in front of the cerebral ganglion; two large nerves *U* issue from it to join the much higher pharyngeal ganglion, this is the pharyngeal commissure.

The pallial nerve *Pa*** leaves laterally from each side of the visceral ganglion. Secondary visceral nerves also issue from behind and below this ganglion, while the large visceral nerve issues obliquely from below it. There are several slender nerves on the sides, which, according to Chun's figures would be the olfactory Y and opthalmic Z nerves, but I could not verify this.

The eye exhibits very interesting features. It is laterally flattened so that from the front its aspect is rectangular (Pl. IV, Fig. 5) with gentle angles and slightly curved sides. It is approximately 10 mm long and 6 mm wide. Viewed from profile it is posteriorly rounded and its contour is almost semicircular. It is very dark brown—almost black. I think this color is due to the pigmented retinal layer because there is no other colored part in the organ. The lens is strongly protuberant and large in relation to the size of the eye (Pl. IV, Fig. 5). The cornea and hyaline connective tissue join it to the skin. The eyelids could not be distinctly seen due to the poor condition of the specimen.

*Should read *GV* as in the plate—General Editor.
**Shown as *P* in Fig. 4 of Plate IV—General Editor.

The remarkable feature of the eye is the short, broad membranous tube stretching from the exterior to the anterior chamber of the eye; at its base it is extended by a circular membrane around half the periphery of the lens. This tube is joined to the skin; it is gaping, shallow and is above and to the left of the lens. I found it comparable to the occular aperture of oegopsids and sepiolids and in this octopod also it has the same relations. It is a vestige of the transformed pseudocorneal slit.

The optic nerve reaches the posterior face of the eye; it bifurcates into two branches (Pl. V, Fig. 7), one ascending *A*, the other descending *D*. Each gives out nerves that spread fan-like over the posterior, curved surface of the ocular globe, following the meridians. The main short branches are slightly swollen before ending in a brush of straight cords (Pl. V, Fig. 7). The ascending branch consisting of two parallel entwined strands joins below the white body *B* applied against the eye and ramifies here.

The optic ganglion forms a yellowish mass *O* at only 1 or 2 mm from the eye. It is noteworthy that this mass is not attached here and the optic nerve is absolutely distinct between it and the base of the eye. The optic ganglion is kidney shaped (Pl. IV, Fig. 6) and slightly twisted upon itself so that its plane of symmetry is leftward. The optic nerve *P* arrives at the hilum of this ganglion which is opposite the brain. The optic nerve swells slightly into a mass at the point of penetration into the ganglion. On the other side, the swollen side, the optic nerve exits the ganglion to arrive at the eye.

The peduncular ganglion *C* is at the same point where the optic nerve penetrates the hilum of the optic
24 ganglion; it is a small, yellow, spherical mass distinctly marked, juxtaposed to but not fused with the optic ganglion. A very small, gray, oval-shaped spot *M* is present on the optic ganglion opposite the point of entry of the optic nerve and in very close proximity to the peduncular ganglion.

Numerous, regularly spaced, curved striations are visible on the surface of the optic ganglion. These issue from the nerve's point of entry, diverge, then reapproaching and reuniting at the exit point of the optic nerve toward the eye. It was not possible to make sections to study its structure.

As already observed, the optic nerve is extremely long (Pl. V, Fig. 6), curved and without a peduncular ganglion on its path since the latter is adherent to the optic ganglion. This is a distinguishing feature since it is one that differentiates the genus *Vitreledonella* from the two other genera.

The sections of the retina (Pl. V, Fig. 8) showed some interesting features. The very long rods rest with their lower tips in a beveled fashion on a lower, very thin, delimiting membrane *L*. A thick, continuous, brown layer is slightly above this membrane in the lower fourth of the rods (Pl. V, Fig. 8); its surface directed toward the inside of the eye is flat, whereas the surface turned toward the delimiting membrane is irregular, winding, and serrated. Due to improper mounting, I could not see how the pigment grains were distributed between the rods. Prolongations of the nerve fibers and the lower tips of the rods are observed between the delimiting membrane and the pigment layer. A layer of nerve fibrillae is below the delimiting membrane. The greater the thickness in which the oval nuclei *A* are embedded, the greater the oblique padding of the nerve fibers *D*. It is here that the fan-like terminations of the optic nerve end in bundles with probably connective tissue nuclei. A membrane *C* acting as the sclerotic layer is below the connective tissue. It is thin, interrupted and not compact.

The retina probably has some folds within the eye but I could not confirm this on the fragment of deteriorated eye from which I took a section. At points where appearances of folds are present, whether natural or due to mounting, the distribution of pigment can be seen with greater detail than where the pigment has been vertically cut. It seems to consist of innumerable extremely fine granules stuck against the tip of the rods and plunged in a transparent, convoluted substance.

It must be finally noted that the rods are not perfectly parallel tubes everywhere; in the folded regions particularly, they swell out apically into more or less large uneven masses so that the largest reach to the internal delimiting membrane *S* and the smallest ones stop at a lower level.

The *white body* (Pl. V, Fig. 7, *B*; Pl. V, Fig. 8, *M*) is pink, flat and slightly thick; it is approximately

25 1.5 mm in diameter. It resembles a small, irregularly-shaped lens and covers the lower branch of the optic nerve at the point of its expansion on the ocular globe. Four or five small masses of pink substance (Pl. V, Fig. 8, *M*), exactly resembling the ocular globe masses, are present along the length of the optic nerve not far from its exit from the brain. They are isolated and embedded in the surrounding gelatinous substance. A few more are disseminated among the viscera close to the heart for example.

The sections show that this white body consists of cells with large nuclei (Pl. V, Fig. 8, *M*) grouped in small masses. The whole forms a raspberry-like surface with small vessels and no secretory duct. This organ does not seem to differ from the similar organ found in other cephalopods except in color and reduction in size. The tube indicated at *V* (Pl. V, Fig. 8) is a blood vessel.

I could not distinguish if a cutaneous olfactory organ is present, or if there are cutaneous papillae near the subocular region as in many other octopods, since this region was too deteriorated.

Mantle cavity (Pl. V, Fig. 9). The mantle is a thick, soft, transparent flaccid pouch, with a gelatinous consistency. Its very wide opening surrounds the funnel. The visceral mass is embedded in an abundance of gelatinous tissue, considerably greater in volume than the viscera themselves. The mantle cavity is divided into two compartments by a vertical, median and ventral, incomplete partition, in which is inserted the terminal part of the intestine and the duct of the ink sac. Toward the upper third, a fine band (Pl. V, Fig. 9, *B*) of transparent tissue attaches the visceral mass to the mantle wall on the right and left; the pallial ganglion is visible in the mantle at the base of each mass. Below on each side, another fine, transparent band in the shape of a long transparent triangle *T* bears the gill inserted by its small base on the visceral mass, one large part is on the mantle the other large part floats free.

Finally, two large, very strong retractor muscles (*R*)* descend from the funnel. They are definitely the two most powerfully developed structures of the mantle cavity in which all other organs are considerably reduced. The retractor muscles arise from the partitions which separate the tubular part of the funnel from the lateral valves and proceed toward the posterior region of the visceral mass where they become lost.

The funnel. This is a well-developed organ. The central thick-walled tube, is moderately elongated; it seems to extend beyond the level of the eyes. The lateral pouch valves are large, thin-walled and soft. Although the entire structure is transparent and gelatinous, the central tube is comparatively firmer. The width of the funnel is in proportion with the large size of the mantle opening.

The funnel has a distinct organ (Pl. VI, Figs. 5 to 7) shaped like a circumflex accent whose angle is directly below the funnel opening. This almost horizontal band is approximately 2 cm long and
26 3 mm wide and is slightly yellowish. This funnel organ thus has a very different shape from that observed in other octopods, and is equally noteworthy for its noticeable lack of development.

The study of its sections, although inadequate because they were not made from a properly fixed fresh animal, has provided some interesting information.

The funnel organ consists of an epithelium (Fig. 6, *E*) of elongated cells enclosed between two adjoining gelatinous layers of connective tissue, a basal dermal layer (Fig. 7, *Gp*) on which the lower tips of the epithelial cells are implanted and a surface layer (*Gs*) through which the upper tips of the cells pass. Small cells are visible here and there in these two layers (Fig. 7, *H*); these are probably the true cells of the two membranes whose mass is formed by the external gelatinous secretion of their envelope.

As for the special cells of the epithelium they have the peculiarity of becoming so strongly impregnated with stain that it is very difficult to distinguish details of their content. These cells are shaped like a capillary tube (Fig. 7, *M*) with a closed base implanted in the basal layer while the upper end gapes open on the surface layer. A generally oval-shaped nucleus (*N*) is in the lower fourth of the cell.

Among these cells, a few are completely filled with a slightly colored flocculent material resembling

*In Fig. 9 of Pl. V, *R* is labeled as the urinary opening. The retractor muscles are not labeled but are obvious—Scientific Editor.

mucus (Fig. 7, *O*). Others are only half filled with similar mucus, the other half being filled with the matter described earlier. The majority of the cells are, however, completely filled with spindles pointed at both ends (Fig. 7, *S*) and made of a transparent, compact homogenous matter, actively staining with hematoxylin. Some cells contain only one spindle, others two or sometimes three. It is the tip of the most external of these spindles that protrudes above the peripheral membrane. This is observed nearly everywhere in sections which have become detached during handling.

It is very difficult to interpret the role of these small organs. They are evidently secreted by the very numerous special cells of this epithelium, but I do not know if they dissolve in the sea water passing through the funnel or are expelled at will by the animal as a defence mechanism. They would thus function like small arrows, probably irritants like nematocysts. In any case they seem to be very easily detached from the epithelium that produces them.

I could not detect nerve cells in this organ.

Mantle viscera. This cephalopod, already unique in its cephalic and brachial structure, becomes more remarkable when its mantle viscera are studied.

As is frequently the case in some pelagic decapods, a considerable reduction of all the organs is found in this octopod. The reduction is such that the viscera occupy only an extremely small space in the large mantle which has a gelatinous consistency. It is odd to observe that the digestive system is so reduced in
27 a body so well developed. The entire digestive system, renal organ, heart, gills, liver and salivary glands have all been miniaturized. The reproductive system is not only reduced but so strongly displaced that it adjoins the digestive organs. All these viscera were probably united into as small a mass as possible to decrease visibility either through making it as transparent as possible, or to decrease in size all strongly opaque organs, or to envelope the opaque organs with a thin membrane containing chromatophores and golden iridocysts. This has already been observed in the nervous system and buccal bulb; we now see it in the genital and digestive systems. This is identical with the phenomena observed in pelagic gastropods such as carinariids and especially in *Firola* [Pterotracheidae].

If the mantle cavity is split open, the organs of various colors are visible. Some, such as the renal organs, heart and genital gland, are opaque due to the action of the preservative but they would be transparent in the live animal.

Digestive system. As mentioned earlier in the text, this system is first of all notable for its small size. Its volume is not one-fourth of that of a coastal eledonids of the same size. Its liver alone is more than 20 times smaller.

However, it is noteworthy that these organs have undergone considerable displacement. While maintaining their basic connections, they have so markedly changed their position that it becomes difficult to find their homologues. The reduction of the bulb and grinding organs that it encloses, the extreme narrowness of the esophagus, absence of a salivary gland, smallness of the spiral stomach and liver and reduced caliber of the rectum are all characteristics that evidently make this cephalopod an animal whose organic feeding can be taken only in the form of very small plankton or disintegrated organic matter.

A. *Buccal bulb* (Pl. V, Fig. 11). A circular pad-like lip festooned with 15 rounded denticles encircled by the crown of eight No. 1 suckers on the arms exposes the edges of the beaks. The entire bulb is very small, laterally flattened and enveloped in a membrane with brown chromatophores with some golden dots. It would be at least ten times larger in a common eledonids.

The musculature of the bulb is very weak.

The bulb has the form of a truncated cone with an irregular base and which carries a large cylindrical papilla like a spear directed toward the base (Pl. V, Fig. 11). The oval, yellow, buccal ganglion lies between this papilla and the esophagus. This entire mass is only 8 mm at its widest point. I could not detect any pharyngeal salivary glands; if present, they must be extremely small.

B. *The beaks* (Pl. IV, Figs. 7 and 8) are weak, only their cutting edge is hard. They are 4 mm at their
28 greatest diameter; the color of the free edges is black, followed by a brown zone, then yellow and finally the external margin which is colorless and transparent. There is not much difference between the two beaks; they are both much flatter than in ordinary cephalopods; the lower one has in its center an extremely small, scarcely visible dot, bordered by two minute depressions. Due to the smallness of the bulb and thinness of the beaks, the muscles must be very poorly developed. These beaks evidently cannot dismember large prey.

C. *Radula* (Pl. V, Fig. 12). Like the other organs of the pharynx, the radula is very small for such a large cephalopod. Its median tooth is provided with 5 points, of which the center one is more developed than the others; the first marginal tooth has 2 points, one large the other small; a second marginal tooth has one point; one small marginal tooth has a long tip; there is a lozenge-shaped bordering plate. This very short radula covers a tiny papilla.

D. *Esophagus* (Pl. V, Figs. 10 and 13). This tube is noticeably long and narrow; it is almost capillary-like. Preservation in formalin has turned it white. It seems to enclose some muscles. It traverses the brain and then becomes flexious before passing into the stomach. I do not know whether this arrangement is natural or due to poor preservation.

A little above the entry of the esophagus into the stomach, there is a slender appendage; it is vermiform, white and bent into a semicircle (Pl. V, Figs. 10 and 13). This represents the crop of other cephalopods, reduced here to a vestige that probably cannot be used for digestion. This organ is usually an accessory to the digestive system that has here been almost completely eliminated in the general reduction of the viscera.

E. *The stomach* (Pl. V, Figs. 13 and 14), is an ovoid pouch with a smooth outer surface, on its walls can be seen lighter lines corresponding to internal ridges. The organ has two parts: the larger one to the left where the two ends of the alimentary canal are attached; the smaller one to the right, separated from the first by a groove. In the larger part there are three inwardly protruding, parallel ridges. In the smaller part, seven of these issue from the fold that separates the two halves of the stomach.

F. *The spiral stomach* [caecum] (Pl. V, Figs. 13 and 14) is completely separated from the main stomach by a short intestinal canal which forms a diverticulum. It is smaller than the stomach and forms a kind of spiral tower. It is ornamented with spiral lines corresponding to almost parallel internal ridges. Its wall seems to be transparent; small, yellowish, oval-shaped spots, which are probably parasites, are scattered about. Two hepatic ducts enter at the base of the spiral stomach, one passing above and the other below the beginning of the intestine.

G. *The intestine* (Pl. V, Fig. 13) is short and becomes increasingly narrower with distance from the stomach. After the spiral diverticulum it makes a horizontal curve, passes between the two hepatic ducts, extends beyond
29 the level of the liver and attaches itself ventrally to the genital gland; it follows it vertically to its tip, continues to make a curve and reaches the membrane of separation from the mantle cavity; it reascends vertically and terminates at the small anal swelling which is provided with two small lateral lamellae. The entire end of the intestine from its exit from the genital gland is extremely slender, transparent, narrow and capillary-like as in the case of the esophagus. The duct of the ink gland terminates in the anal ampoule.

The gastric ganglion rests at the angle formed between the stomach and the spiral stomach [caecum], against the small part of the intestine between these two pouches. The nerves leading to the entire central portion of the digestive apparatus issue from this ganglion (Pl. V, Figs. 10 and 14).

The entire stomach, the first part of the intestine and the bases of the liver and genital gland are enveloped by a thin membrane that follows their curves. It is full of large yellow and brown chromatophores and is strewn with numerous golden iridocysts.

H. *Salivary glands* (Pl. V, Figs. 10 and 13). The two large abdominal salivary glands of the eledonids and other octopods are here reduced to a single tiny, lanceolate gland at the very end of the esophagus near the rudimentary crop and applied against the top of the stomach. This is very different from the normal position. A very thin duct issues from this small gland; it runs along the entire length of the esophagus and, after traversing the brain, penetrates the buccal bulb. The gland is light brown; its duct is colorless and transparent.

I. *The liver* (Pl. V, Figs. 10, 13, 14). This organ is also extremely small and the position is reversed in relation to the normal situation. Since the stomach is very high, there is no space for the liver between it and the head. Moreover, the liver is turned and is below the stomach. It resembles a small pink stalk closely united with the ink sac which is almost as large and envelopes a large part of it. It covers a large portion of the spiral stomach and even reaches the stomach so that the upper part of the gland has to be cut to detect its connections. This dissection is extremely difficult in view of the small size of the organs and their delicateness.

Two normal hepatic ducts issue from the liver and encircle the intestine which passes between them (Pl. V, Figs. 13 and 14). The liver is pink, opaque and enveloped in the gastric region by the membrane with chromatophores and golden iridocysts which has already been discussed.

If the upper part of the liver is moved aside, the two united hepatic ducts can be seen as they penetrate the spiral stomach. One of the ducts expands slightly into a small swelling (Pl. V, Fig. 14). They diverge and penetrate the hepatic tissue. It is not known whether one part of the hepatic gland performs the function of the pancreatic gland.

A study of the sections shows that the hepatic tissue consists of polyhedral cells that do not take hematoxylin stain although their nuclei become deeply stained. The cells are grouped around the canaliculi and constitute a network of entangled tubes (Pl. III, Fig. 1, *G*).* Capillaries in the tissue occasionally contain spherical corpuscles with nuclei in the shape of a C, an S, spiral, in grains, etc. (Pl. VI, Fig. 1, *G*). They do not resemble the nuclei observed in other organs, perhaps they are in the process of multiplication. A lacuna is observed under the connective tissue membrane around the liver. This lacuna is traversed by several connective tissue trabeculae with many large slender, reddish, lymphoidal cells. They will be discussed further in the text in connection with the gills.

The long ink gland is embedded in the thickness of the liver (Pl. III, Fig. 1, *N*). In a thin section, its edges are largely hidden by the hepatic tissue. I again stress the noteworthy fact that this liver is much smaller than that of other cephalopods of the same size.

K. *The ink sac* (Pl. V, Figs. 9, 10, 13). The gland itself consists of a long black cylinder embedded in the liver for almost its entire length. It begins slightly below the stomach and descends to the end of the liver (Pl. V, Fig. 10 *M*). It folds on itself to form a tube or reservoir *N* which again ascends vertically along the length of and outside the hepatic cylinder. The result is that the liver is squeezed between the two branches of the ink tube. The reservoir is detached from the liver slightly higher up; it becomes free and is then transformed into an extremely long, small, thread-like canal which passes behind the liver then above the intestine and forms a ring to recross the liver and ultimately reaches the mantle. It reascends vertically and discharges into the anal ampulla through a capillary tube. Small swellings like beads of a rosary, each a small supply of ink, are along the entire length of the ink ejector tube.

The edges of the ink sac touching the liver are screened by the pink elements of the latter. The glandular hepatic structure can be easily seen against this black background.

The entire ink sac is covered by a layer of fine golden iridocysts enclosing it in a metallic-like sheath.

This general arrangement of the ink organ is extremely interesting. The basic relationship is maintained, i.e., the opening in the rectal ampulla, which is the embryonal beginning of the gland, fuses with the liver. However, the rotation of the position of the liver and its displacement toward the top of the stomach have resulted in an anatomical modification of the ink gland. Its secretional section, rotating itself with the liver, is directed downward from the top and, in order to regain its normal position, the reservoir is folded into a *V* along the length of the liver. The torsion of the liver, which not only changed direction in the vertical

*Not labeled in drawing, although mentioned in explanation—Scientific Editor.

31 plane but also made a spiral turn on itself, has forced the ink duct to make the same torsion; this is definitely the cause of the bead-like ring in the ejector canal. I do not know of any other cephalopod with an ink sac that has the same complicated arrangement. Moreover, the secretory part of the ampulla is at the base, the reservoir above, and the generally short terminal canal is on top rising above the ampulla in the vertical and median plane of the animal.

Respiratory system. As in other cephalopods, this form breathes with two gills. However, here again, these organs are very greatly reduced and are approximately only 1 cm long, i.e., a minimal size in relation to the size of the animal. Additionally, each gill is only half the gill (demibranch) of a normal octopod. It is known that in these animals this organ consists of a circular series of arcs resting on a branchial gland, like the wooden ribs on the keel of a boat. But in this animal half of each arc is missing and the gill is only a concave plate on which rest the arched respiratory pads between which water passes through the slits. Only the lower part of each circular arc remains; the upper half is represented by three small remnants which, in view of their size, can only play a very minor part in the respiratory function.

The gill is supported on each side by a thin, hyaline, elongated, triangular lamella fixed at its base to the visceral mass and by its lower edge to the mantle. The upper edge floats freely in the mantle cavity. The gill is attached toward the base of this transparent lamella. It should be mentioned in passing that this lamella is the same member which, in ordinary cephalopods serves as the ligament for the gill. However, when the gill is large the ligament is partially hidden and seems to be totally accessory. It is the opposite in the animal under discussion, the ligament is three times longer than the gill which is very small and appears to be an unevenness in its base.

Because of the oblique position of the ligament in the mantle cavity it might be thought that the gill is on its lower side where it consists of protuberant pads. On the contrary, its upper concave face contains the opening of the slits and the rudiments of incomplete pads. The respiratory parts of the gill are pale pink and consist of tufts of filaments or closely packed lamellae.

There are seven branchial arches; the largest at the base, diminishing in size to the tip on the lower face of the lamella. On the free ridge there are three small masses of pink tissue representing the pads of the upper arches that have disappeared or atrophied. Only these seven pads are observed on the lower face. They are wavy, pink, protuberant and separated from each other by slits for the passage of water. Things are more complicated on the concave upper face; each arch is supported by a transparent, hollow, semi-
32 cartilaginous tube issuing from the branchial crest to reach the base. But this arch is not straight, it is wavy, larger at the summit than at the base. These arches are not initially parallel; they diverge slightly like the fingers of a hand. Fig. 16, Pl. V shows the arrangement better than would a lengthy explanation.

The base of each of these arches can be seen in the hyaline tissue where the gill is inserted. The vessel issuing from the branchial vein is embedded here as in a small funnel.

It is the afferent vessel that supplies venous blood to each arch. It distributes the blood in the pink respiratory tufts where hematosis takes place. This blood then passes into the efferent vessel parallel with the afferent vessel but not inside the cartilaginous arch. It is more superficial below the plaited pad of the arch. The seven afferent vessels issue from the vein and the seven efferent reach the branchial artery which carries oxygenated blood to the heart. Pl. V, Fig. 16 diagrammatically depicts this gill arrangement.

The section of a branchial arch (Pl. VI, Fig. 2) shows the fibro-cartilaginous nature of the hollow support (*F*) through which the venous blood arrives. The branches of this cartilage act as a support for the wavy epithelial folds of respiratory tissue and continue to enclose the distributory channels of venous blood. Toward their base these convoluted folds contain afferent canaliculae recognizable by the round corpuscles of blood and nuclei that they enclose. The periphery of these folds is occupied by afferent capillaries protruding at the surface of the gill (Pl. VI, Fig. 2, *A*).

The pink color of the gill is due to the very large number of red dots (B) on the thickness of the wavy tissue. They are also disseminated in smaller numbers in the white tissue of the branchial gland. At high magnification, these elements appear to be red cells with a dark nucleus widely varying in size; some are enormous, full of granulations and lodged only in the surface capillaries. This red color is seen in an animal that is not artificially colored. It is not known whether they should be considered lymphocytes with respiratory hemoglobin, or cells with an excretory function. Similar cells have already been observed in the liver. They are always at the surface of the organs, never deep. According to the studies on other cephalopods by M. Cuénot, they are probably organs charged with the function of eliminating the toxic products of the blood. Here and there, and also on the surface, are bunches of cells resembling nerved endings. It is impossible to define all these histological details in the absence of mounts from a fresh animal.

The branchial gland, called the *spleen*, common in all cephalopods, is observed along the entire length of the gill and is included within the thickness of the support lamella. It is extremely thin, transparent, colorless and difficult to distinguish. It is strewn with the red dots that have just been discussed for the gill. It is probable
33 that it receives venous blood from each of the vessels that accompany the seven arches and from the vessel extending the gill at its tip. However, not having been able to make injections, I cannot prove this. This gland curves in an arc at its point (Pl. V, Fig. 15, *G*), which gives it proportionately greater length than in other octopods.

The sections (Pl. VI, Figs. 3 and 4) show that this organ is a flat strip included in the connective tissue of the support lamella of the gill. The thickness rarely includes more than around ten rows of cells. The gland is pitted with curved, irregular lacunae (*L*) that contain blood corpuscles. The true cells (*C*) of the organ are polygonal, irregular and have vacuoles. The fixed brown cells (*D*) are the same as those seen in the liver and gill. A larger venous sinus occupies the lower and upper edges of the organ.

Due to the action of formalin the branchial nervous system has become opaque, white and is clearly seen in the transparent tissues. The main branchial ganglion is at the base close to the lower point of the branchial gland. It is the first in a series of beads running between the base of the gill and the spleen. This ganglion curves at its tip and is directed toward the three incomplete arches on the free ridge. The beaded ganglia are numerous—at least around 20—but small, indistinct and irregular. A nerve issues from the basal ganglion, and ascends with another small ganglion to where another large nerve rises toward the large vascular trunks (Pl. V, Fig. 15).

The branchial heart is between the base of the gill and the renal organ (Pl. V, Fig. 18). It is rounded, pale pinkish yellow and is traversed by the afferent vein of the gill.

The poor development of the gills does not seem adequate to insure the respiration of such a large cephalopod. It is not even one-tenth of the respiratory apparatus of a coastal eledonid of the same size. One is therefore justified in thinking that the great softness of the tissues permits a supplementary cutaneous respiration. It must, however, also be said that the heart and the main vessels are very small and vascularization is very weak, I hardly found any arterioles. The conclusion is drawn therefore that respiration, like alimentation, is very restrained. The animal is not muscular, is evidently a poor swimmer, has no organs of defense and has a very poor digestive system. It must breathe as poorly as it eats and must be an animal that moves idly in the midwaters or without energetic movement on the soft mud of a bottom that presents no obstacles.

Circulatory system (Pl. V, Fig. 17). No injections were given. Therefore, the observations on the circulatory system are very incomplete.

The heart (Pl. V, Fig. 17) is formed from a ventricle in the form of a yellow, very small lenticular vesicle behind the upper part of the kidneys. The ventricle by itself forms the heart. The two afferent arteries *H* from
34 the gills can hardly be considered auricles. The gills dilate very slightly before reaching the ventricle; they are transparent and do not seem to be more muscular than the other vessels. The aorta *A* ascends vertically with a first branch curving towards the base *U*, and proceeding to the renal sacs and perhaps to the lower part of the body and mantle. The aorta reaches the nervous system which it traverses, as already mentioned, and one of its branches is attached to the axial nervous system of each arm.

A small artery *E* issues from the ventricles and seems to proceed toward the liver and the ink sac. Another artery *C* initially proceeds downward and then ascends. I followed it to the genital gland.

Regarding the venous* system, its central part consists of two parallel veins descending to the renal organs (Pl. V, Fig. 18). Slightly above this area they seem to become anastomosed. They then penetrate these two glands, pass through them and emerge as the afferent branchial vein. They pass by the branchial hearts, pink, opaque masses as large as the ventricles, before penetrating the gills.

Slightly above the stomach, the two large veins abruptly narrow almost to capillaries before passing through the nerve collar. In the arms they form an artery-like vessel joined to the axial nervous system (Pl. V, Figs. 2 and 3). As far as can be judged, there are very few capillaries and small vessels in the thickness of the mantle, cephalic or brachial walls. Circulation is definitely very restricted.

Excretory system (Pl. V, Fig. 18). Like the venous system mentioned above, the excretory system occupies the normal position of other octopods, except it is much smaller. The urinary sacs are very poorly developed and are separate except in the upper part. Each opens out through an aperture above the gill. This aperture is on the tip of a transparent papilla (Pl. V, Fig. 18, *R* and Pl. V**, Fig. 19), emerging from a pit. The branchial efferent artery passes above the base of this papilla before becoming an auricle.

The renal gland is comprised of two opaque, white, cylindrical superimposed parts *D* forming an elbow. It consists of convoluted tissue but is not villose as in other octopods. The gland seems to almost fill its urinary sac *U* which is still less developed than the renal gland itself.

Genital gland (Pl. V, Figs. 10 and 19; Pl. IV, Fig. 9). The specimen is an immature female. The system consists of an oval-shaped ovarian pouch with the tip applied to the stomach. It descends vertically almost to the middle of the hepatic cylinder (Pl. V, Fig. 10). The initial portion of the intestine closely adheres to the ovarian gland.

The ovary is a soft-walled sac (Pl. V, Fig. 19) with two layers: an outer hyaline connective tissue layer and an inner spongy layer with small yellowish filaments carpeting the entire sac. However, in the left middle region there is an internally protruding lump (Pl. V, Fig. 19) which is the gonad. If a fragment is detached, it can be seen that it consists of a fibrous connective tissue network mixed with hyaline tissue containing occasional very small, orangish yellow, oval-shaped, polynucleate eggs. There are small red cells forming a network; these are probably the same as those in the gills.

The sac is enveloped by a thin membrane which, as described for the stomach, contains chromatophores, but these only cover the side of the ovary where it can be seen from the outside. They are absent on the side joining the liver or intestine which is not visible from the outside as it is masked by these organs. These chromatophores are elongated and arranged to form violet horizontal streaks. Fig. 19 shows them on the right hand side where the dissected membrane is moved aside. They can be seen inside the ovarian sac through the transparency.

The cavity of this sac is basally contracted like a funnel and immediately bifurcates into two canals which are twisted into a spiral encased in a hyaline sheath. The canals themselves are so transparent that it is extremely difficult to follow them. They reach to the oviducal gland where they form an S-shaped ring and reach the outward opening terminal papillae (Pl. V, Fig. 19)†.

This opening for laying is reduced to an oval-shaped slit not much larger than the tube leading to it. It is next to the urinary papillae (Pl. V, Fig. 19)†.

The oviducal gland (Pl. V, Fig. 20) has the form of a biconvex lens intercalated on the oviduct which reaches the gland and issues from it almost in the middle of the two convex faces. This gland consists of a large number of small, elongated, pyriform glandules whose ducts all converge toward the outlet of the

*Misprinted as "Nervous" in the French original—General Editor.

**Plate IV in original—Scientific Editor.

†In the original this is given as Pl. IV, Fig. 9—Scientific Editor.

oviduct. The organ resembles an orange whose segments correspond to the glandules and the secretory canal to the peduncle of the fruit. This structure is distinctly visible because the connective tissue separating the glandules is absolutely transparent. The glandules have become white and opaque due to the action of formalin and are striking due to the clarity of their shape. They are approximately 4 mm in diameter.

The histological examination of this gland shows some interesting details. Each glandular constituent (Pl. VI, Figs. 10 and 11, *Gc*) is S-shaped with an ampulla-shaped middle region forming the main secretory part.

The secretory canal is also lined by rather elevated epithelium and probably also aids in secretion. In the lower reservoir part, the epithelium (*R*, same figures) is almost flat, more rare and clearly non-secretory. The oviduct entirely transverses the gland, encased in circular and longitudinal fibers which are more closely packed as they approach its epithelial wall. The reservoir is filled with convoluted transparent strands of mucus
36 not stained by hematoxylin. The elementary glands are embedded in the lacunal hyaline conjunctive tissue. All their ducts discharge at almost the same level, at the exit of the oviduct from the oviductal gland (Pl. VI, Fig. 12).

Summary and conclusion. This cephalopod has been studied as completely as possible with a single female specimen in poor condition.

It is extremely interesting because it is the farthest removed from the normal type species of octopods known to date. It represents an adaptation to the bathypelagic life of a form affiliated to the family of Eledonidae by the single row of brachial* suckers but greatly differs from it in numerous points.

The most striking difference is the over-development of the connective hyaline tissue and the reduction of the muscles and all the viscera. In an animal of a size comparable to that of an average coastal eledonid, one would expect to find viscera of the same proportion, but this is not the case since the size of the animal is approximately 30 cm while the organs are no larger than those of an eledonid 5 cm long. Similar facts have been observed in various bathypelagic decapods, for example a few cranchiids but I am not aware of reports of any freatures approaching this in the octopods.

It must be noted that these small viscera were significantly displaced as well as reduced. This has resulted in radical modifications in the disposition of the digestive tube, liver, ink sac and genital gland. These inverted organs are lodged very high under the funnel, and are enveloped in a membrane containing chromatophores and golden iridocysts which disguises the entire small and opaque visceral complex. On the contrary, organs such as the kidney, heart and genital ducts, which must have been transparent in the living form, do not have this colored and iridescent membrane.

The buccal bulb and central nervous system are both very small and are enveloped in the same manner.

The flattened eyes have a tubular duct through the pseudocornea as in the decapods.

The anatomical and external features are so unusual that I am justified in creating a new genus for this cephalopod. I name it *Vitreledonella* in reference to its affiliation with *Eledonella* and its vitreous transparency. The fundamental traits of its nervous system and optic ganglion differentiating it from genera *Bolitaena* and *Eledonella* have been discussed earlier in the text. It further differs from them by the greater length of the arms, the enormous displacement of the eyes and the softness of its tissues, especially that of its arms which are almost without muscles. It would be redundant to mention all the distinguishing anatomical characteristics enumerated in this section, markedly differentiating it from all the other cephalopods.

37 Some conclusions on the systematics of the family Eledonidae can be drawn from this study and earlier publications. At the present stage of knowledge this family includes all octopods without fins and with a single row of suckers. These are the only fundamental characters uniting the various genera of Eledonidae.

However, this affiliation seems to be rather weak in view of the enormous differences existing between them. While admitting that the bathypelagic forms such as *Vitreledonella* may have been derived from

*In the French original given as "branchiales" which is an obvious misprint—General Editor.

coastal forms like normal eledonids, they now differ so much that they can no longer remain together under the pretext that they possess a single row of suckers. However, their separation would be premature at this stage in view of our inadequate knowledge of the anatomy of the majority of the species.

Along with the normal eledonids, a separate family must be erected for the bathypelagic eledonids by accepting the name Bolitaenidae of Chun. This will surely be the same for genus *Moschites* (old *Eledone*) which may be divided into two genera, if only on the basis of the great difference of the hectocotylus in the various present species. One genus groups together the species with a smooth skin and thread-like hectocotylus living on the coast (*M. moschata, M. cirrhosa*) and the other for the benthic and polar forms with a rough skin and large hectocotylus as in the *Octopus* (*M. verrucosa, charcoti, challengeri*, and *media*).

They differ sufficiently even if only by these features, to be separated into two genera: *Moschites* and *Graneledone*. Unfortunately, we do not know the anatomy of the latter.

Based on our present knowledge, I give below my opinion on how the eledonids should be classified.

Family ELEDONIDAE

Octopods with a single row of suckers and no fins.

A. 1st subfamily Eueledonidae

Normal forms, muscular, opaque and non-pelagic.
1. Littoral or shallow seas species with smooth skin and filiform hectocotylus.

Genus *Moschites*

(*M. moschata*, Mediterranean; *M. cirrhosa*, North Altantic).

2. Benthic or polar species with rough skin and large hectocotylus.

38 **Genus *Graneledone***

Benthic species: (*M. verrucosa* Verrill, *M. challengeri* Berry and *M. media* Joubin).
Polar species: (*M. charcoti* Joubin and *M. turqueti* Joubin).

B. 2nd subfamily Bolitaenidae Chun

Bathypelagic adaptation, body more or less transparent.
1. Forms with condensed central nervous system, little different from normal eledonids.

Genus *Bolitaena* Steenstrup
(*B. diaphana* Hoyle)

2. Central nervous system displaced by the spacing of the eyes.
a) Peduncular ganglion in the middle of the very long optic nerve (see Chun).

Genus *Eledonella* Verrill
(*E. pygmaea* Verrill)

b) Peduncular ganglion attached to the optic ganglion, very long optic nerve without a ganglion along its length.

Genus *Vitreledonella* Joubin

(*V. richardi* Joubin; *V. alberti* Joubin)

Vitreledonella alberti, Joubin

(Pl. II, Figs. 1 and 2)

Expedition of 1911: Stn. 3128, depth from 0 to 3,500 m. August 15, 33°40′ N to 33°52′ N; 19°W to 19°16′ W. Net with three layers. Northwest of Madeira. Stn. 3131, depth from 0 to 3,500 m. August 16. 35°09′ N; 21°21′ W. Bourée net rapidly retrieved. Between Madeira and the Azores.

These two cephalopods, whose habitat appears to be abyssal, resemble on the whole the preceding species, *Vitreledonella richardi.* But careful examination of these specimens provides definite indications that they differ in several details. I have, therefore, decided to erect a new species for them: *Vitreledonella alberti.* These species are placed close to one another in this extremely interesting genus.

39 The following description refers exclusively to the larger specimen. I shall provide some complementary information on the second which is in poor condition. However, it is interesting to observe that these two specimens differ greatly in size but are almost identical, negating the hypothesis of great changes with age.

The animal has a characteristically squat body with generally transparent tissues turned somewhat milky due to storage in formalin and generally soft organs which give it a consistency like a medusa.

The length of the animal, including the arms, is approximately 14 cm; when alive, it was perhaps slightly longer. The body is saccular and is approximately 4 cm long and wide.

The head is scarcely discernible between the bases of the arms and the commencement of the mantle. It seems to be reduced to the location of the large eyes and the narrow band of hyaline tissue uniting them dorsally. It must be mentioned here that in the illustration on the left of Pl. II, representing the dorsal side of the animal, the edge of the mantle is continuous on the dorsum in a distinct line; this is not a dorsal slit, as one might think, but a thin fold due to the contraction of the animal and affects only the skin. The space between this fold and the bases of the arms and, laterally, between the mantle and the two eyes shows how much the corresponding space for the head of this cephalopod has been reduced.

The distance between the two eyes on the dorsal side of the head is 21 mm, which is considerable for the size of the animal. These eyes are distinguished by their elongated shape and the position of the two lenses. Their position permits viewing objects above the dorsum and slightly to the side, but not below the animal.

A superficial examination of the eyes shows some interesting features. Its greatest diameter, transversely directed, is 10 mm. However, its nuchal end [dorsal] is larger and more rounded than its funnel end [ventral] which is more pointed. The lens is set in the middle of the eye. A golden, iridescent ring surrounds the entire circle of insertion. The pointed funnel end of the eye has the same color whereas the swollen nuchal end is black.

The entire eye rests on a large, short, transparent stalk through which a large bilobed optic ganglion can be seen; thus, it is very far removed from the brain to which it is connected by the large optic nerve that can be easily detected due to the transparency.

The funnel is short, large and squat; numerous muscle fibers can be seen through the skin. The two lateral pouches of the funnel are as well-developed as are the two muscle pillars. This funnel is apparently
40 adapted for powerful swimming, and thus compensates for the inadequacy of the arms for this purpose. The orifice for the outlet of water from the funnel is transverse, bordered by a bilobed membrane with a flexuous end. When these two kinds of lips are separated, one can see a protruding fold on the floor immediately behind the lower flap. Since I did not wish to damage the specimen further I was not able to detect any structures on the roof of the funnel.

The mantle cavity is divided into two symmetrical halves by a strong partition inserted under the funnel and in which the anal aperture is located. I could not detect whether the partition is complete down to the base of the mantle cavity; it is so at least in the upper half.

The viscera within the water-filled mantle cavity form a fusiform, elongated, vertical but slightly oblique mass, in the median chamber. Their ensemble is very much reduced in relation to the overall size of the body. The gills are extremely small, smaller than in *V. richardi.* They give the impression that they are there merely to conform to the morphological rule, but physiologically do not have an important function. Their respiratory function must be almost completely accomplished by the skin whose fineness and delicate structure must provide considerable permeability for gaseous exchanges.

It was not possible to detect the disposition of the reproductive organs or even to determine the sex of the animal. No indication of a hectocotylus was detected on any of the arms. Some information on the suckers will be supplied later in the text.

On separating the arms, a group of circular membranes is detected surrounding the mouth. The mouth is very narrow revealing the tips of the black beaks. Around the opening is the circular lip whose radial folds are well-marked. Surrounding this, a second circular lip forms a protuberant pad, without ornamentation or bulges. This entire organ seemed to be weak, small and out of proportion to the size of the animal.

The arms are shorter and more depressed than in *Vitreledonella richardi*; they are transparent and only decrease markedly in diameter at their extreme tip.

They are interconnected by an extremely delicate, transparent membrane traversed by an irregular network of fibers. Unfortunately, the web is so distorted that I could not clearly establish its shape. My findings are reproduced here. The membrane between the two dorsal arms extends to the level of the third sucker; it apparently continues as a small crest to the middle of the internal margin of the two arms. This membrane is thus a
41 dorsal membrane and is a continuation of the skin of the head. There is no web between arms one and two. A membrane as fine and delicate as the preceding one is between arms two and three and between arms three and four. They apparently end at the level of the fourth suckers and continue laterally as rapidly attenuating crests. A transverse membrane above the funnel is observed between the two ventral arms. It is short and probably does not extend beyond the level of the second sucker.

The arms of the larger specimen were measured and their lengths (mm) are:

1st arm (dorsal)	..	97
2nd arm (lateral)	..	96
3rd arm (lateral)	..	95
4th arm (ventral)	..	90

It is clear that these arms are almost equal in length except for the fourth which is distinctly shorter.

The suckers have a characteristic arrangement. The first sucker of a series of eight on each arm is very close to the mouth. Together the eight form a circle around the lip. These suckers are hardly one millimeter in diameter. They are not included in the brachial series. The suckers are approximately 1 cm apart on each arm. On the first arm, the first five suckers grow imperceptibly from 1.5 to 3 mm, the sixth sucker is 4 mm, the seventh is 5.5 mm, the eighth is approximately the same and the ninth is 5 mm. From there a series of 13 suckers is observed, the first two are 3 and 2 mm in size; the others are very small. The last are even microscopic and all the suckers of this terminal series are very close to each other.

On the second arm, the ninth sucker is the largest, more than 6 mm, and is the largest on all the arms. The eighth is almost equal and the others are spaced as in the first arm.

The overall arrangement is the same in the third arm but the most developed suckers are smaller than on the second arm where they attain maximum size.

This decrease in size is further accentuated in the fourth ventral arm. The arrangement and distribution of suckers is still the same but the eighth and ninth are still less developed than in the third arm and, thus, very distinctly smaller than in the second arm.

The opening of these suckers is quadrangular due to the smoothened edges of the organ. Each sucker is at the top of a transparent, highly developed nipple.

I assign a second, unfortunately badly damaged, specimen to this species. The arms are almost completely
42 missing so that it is not possible to give the size of the animal. The mantle is approximately 2 cm long; all the organs are here grouped in an elongated mass curved below like the letter J and entirely enveloped in a golden membrane. The gills are so small and transparent that it is difficult to find them. They are 2 mm long and comprise six wavy segments plus a rudimentary one at the tip. The vertical partition dividing the branchial sac is torn; it seems to have been complete for its entire height.

The membrane of the eyes is golden as has been indicated for the larger specimen.

Below the funnel there is a V-shaped organ with its tip behind the upper lip of the outlet orifice. The two branches diverge and their ends are rounded. The surface of this organ is folded; it stands out slightly in the funnel cavity.

The anus is located in the median vertical partition; it has two lateral filiform appendages. The ink sac is apparently very small with a capillary-like duct.

Genus *Bolitaena* Steenstrup, 1859

Bolitaena (*Eledonella*) *diaphana* Hoyle, 1886
(Pl. III, Figs. 2–16; Pl. VII, Figs. 1 and 2; Pl. XII, Fig. 9; Pl. XIII, Fig. 1)

1885. *Japetella diaphana*, Hoyle (**23**), I, p. 232
1886. *Eledonella diaphana* Hoyle (**24**), p. 107
1910. *Bolitaena diaphana*, Chun (**13**), p. 20

Expedition of 1911: Stn. 3082, depth from 0 to 4,000 m. July 28, 34°20′40″ N, 10°05′ W. Bourée net rapidly retrieved. Southwest of Gorringe Bank, west of Gibraltar. Stn. 3086, depth 0 to 4,300 m. July 29, 33°46′ N, 10°44′ W. Bourée net rapidly retrieved. Locality slightly more southwest than the preceding. Stn. 3128, depth 0 to 3,500 m. August 15, 33°40′ N, to 33°52′ N, 19°00′ W to 19°16′ W. Three-layered net. Between the Deserta Grande Island and Santa Maria (Azores). Stn. 3131, depth 0 to 3,500 m. August 16, 35°09′ N, 21°21′ W. Bourée net rapidly retrieved. Locality close to the preceding.

Expedition of 1912: Stn. 3205, depth 0 to 3,000 m. August 6, 31°01′ N, 23°45′ W. Bourée net rapidly retrieved. Locality close to Madeira. Stn. 3215, depth 0 to 2,000 m. August 8, 30°45′30″ N, 25°47′ W. Bourée net rapidly retrieved. Locality close to Madeira. Stn. 3217, depth 0 to 1,000 m. August 8. Same locality as the preceding. Bourée net rapidly retrieved.

Geographic distribution. This is an exclusively abyssal species and to date has not been found in shallow waters. The known specimens come from the following localities: north of Papua, Marshall Islands, Galapagos, Acapulco, Cape Verde Islands, Azores, Madeira and Atlantic Morocco.

43 It is useless to describe the external features of this cephalopod since several authors have already done it. I shall describe only a few of its anatomical peculiarities.

All the specimens of this species are rather small. None has attained the size given by Chun, which he considers an adult. They vary from 3 to 6 cm, including the arms.

General disposition of the viscera. The dissection of these small cephalopods is extremely difficult because of their softness. When the wall of the dorsal side is lifted, the golden appearance of the peritoneal membrane that envelops the visceral complex is immediately noticeable. Granulations with a metallic sheen are abundant mainly in the anterior part of this complex; they diminish toward the base. A layer of large red and brown chromatophores of various hues is distributed in this membrane and envelopes it in a continuous layer. The buccal bulb is also enveloped in a membrane of red chromatophores.

It is observed that the viscera and brain are enclosed in a large chamber or sinus filled with blood. It has two large dorsal cylindrical prolongations descending to the base of the mantle. They curve inward at their tips and almost meet on the dorsal line. The brain, optic ganglia, esophagus and buccal bulb are immersed and as if suspended in the liquid filling this large chamber. The golden dorsal membrane enveloping the visceral complex is exposed in the gap between the two lateral dorsal sinuses. Conversely, the lateral parts of this same visceral complex are visible in the sinus which covers them. In Fig. 1 of Pl. XIII one of the sinuses has remained intact so that the golden membrane is not visible whereas the one on the right side is open which allows us to distinguish it (Pl. XII, Fig. 9).

This arrangement is a step toward complete transformation, in view of its adoption to pelagic life; we have observed its ultimate in *Vitreledonella richardi*. Here, we see a reduction of the visceral complex, the envelopment of this complex in a golden membrane, the semitransparency of the tissues and the appearance of well-developed chromatophores on the pharynx, brain and the visceral complex. A tendency toward a reduction of the gills is also found. All these features tend to distance it gradually from the typical bottom-water eledonids abundantly represented along our coasts by *E. cirrhosa* and in the polar and abyssal fauna by species with a granulose skin as in *Graneledone verrucosa.*

A few other interesting points were observed on dissecting the animal. The abdominal salivary glands are well-developed and attached to the top of the visceral complex. The liver is very large and completely screens the stomach and crop. The kidney is poorly developed and lies at the base of the complex with the heart, which appears to be terminal. The ink sac is long, narrow, and terminated by a long duct discharging into the anal pouch.

44 *The central nervous system* presents conditions particularly favorable for its study. It is isolated in a transparent fluid and the cephalic cranium is almost completely absent. The details of its organization are clearly visible and the path of the nerves can be traced without much difficulty.

On removing the membrane with the chromatophores which covers the brain, one is struck with the extreme complexity of the cerebroid ganglia (Pl. VII, Fig. 2 and Pl. XIII, Fig. 1). It is dented with symmetrical and protuberant convolutions separated by furrows of varying depths. A group of elongated protuberances is on each side, undoubtedly giving rise to various branches of the optic nerve (*O*). At least four superficial branches (*A, B, C* and *D*) can be detected. More may be present at greater depth. The two pads meet almost at the middle of the brain (*C*) above the esophagus. Due to its curvature, the optic nerve (*O*) seems to be too short in this illustration. Its proportions can be better observed in Fig. 2 of Pl. VII.

A complicated center of small lobular circumvolutions (*M, N, R, S*) is observed in the middle of the brain. These are united to each other by a cord forming almost three-fourths of a circle. The upper part of the brain is formed of two large pads, *U* and *V*, constituting the anterior lobe of the cerebroid ganglion. All these parts are like the setting of a ring through which passes the esophagus (*OE*), the aorta (*AO*) and two veins (*T*). The salivary duct is behind them. The visceral ganglion (*Z*) forms the roof of the organ. The pallial nerve (*Pa*) and olfactory nerve (*Ol*) issue from it. Finally, the peduncular ganglion (*G.p.*) is joined to the optic nerve at its point of entry into the optic ganglion (*X*).

This complex cerebroid ganglia is noteworthy. It attains this degree of complexity in no other cephalopod. The thinness of the membranes covering the brain and the absence of cartilage permitted this study which provided precise information on the locations of the cellular groups. It was also possible to trace the nerve beginnings. The size of the large lateral bundles, out of which the optic nerves issue, is also noteworthy. These will pass through the optic ganglia which are also very large. Two peduncular ganglia cover a large cord separating the optic ganglia and optic nerves and consequently the brain. As far as can be concluded from inadequate evidence, these peduncular ganglia are the centers for the coordination of chromatophore movements. They are in direct relation with the corresponding eye through the intermediary of the optic ganglion and with the brain through the optic nerves. A study would have to be done to detect at which point of the brain makes a connection with the pallial nerves which themselves emerge from the visceral ganglion. They definitely do not originate here, as I saw in embryos of another species.

Auditory organ or statocysts. On dissecting the head, two large transparent vesicles are immediately
45 recognizable at the white, contorted ridge which is contained within their wall (Pl. VII, Fig. 1). These are the statocysts. They are as large as the brain; they meet almost at the ventromedian line, leaving only a narrow vertical band of the brain between them; they border it for much of its distance laterally, reaching and covering the optic ganglia. All these parts of the brain can be distinguished because of their transparency. They are separated from the eyes by only a thin, non-cartilaginous fibrous capsule which replaces the cranium (Pl. III*, Fig. 16, *D*).

The capsule of these statocysts is extremely fragile and almost completely spherical. It seems to be supported by the white filament which is more rigid, of a cartilaginous consistency, and S-shaped. This is an extremely complicated ridge, protruding into the interior of the vesicle. The two ends terminate in a pointed tip. One is at the level of the place where the nerves of the arms issue from the brain. The other, at the opposite pole, is at the level at which the pallial nerves emerge.

On the surface of the statocyst two spots are also observed; one is curved and narrow in the shape of a thin leaf (Pl. III,* Fig. 13) and is above the eye. It is linked to the middle of the S by a nerve to which it is joined. The other is rounded, whitish and attached to the transparent membrane like a small disk. A short nerve enters here; it issues from the lower middle face of the brain; this spot and this nerve have no relation with the S-shaped ridge. The upper end of this ridge receives a nerve from the brain not far from the nerve that reaches the round spot.

This entire statocystic vesicle is formed by a thin, transparent, connective tissue membrane internally lined by a flat, pavement-like epithelium in which the cells do not seem to touch each other everywhere.

I have some reservations regarding the histological structure of this organ, since the mounting was not made from a live animal. At the time of capture, the entire animals were put into sea-water which was slightly treated with formalin. This is inadequate for mounting delicate tissues like nerve endings. My findings will therefore have to be confirmed by better preserved material. However, I do not think I will stray too far from the truth in the following description.

Let us first study the ridge: its composition is very difficult to decipher due to its curves that give very complicated sections oriented in all directions. To ascertain this structure it must be assumed that the organ is arranged in a straight line, vertically spread and sagittally sectioned along its entire length. This has been presented in Fig. 2 of Pl. III. Various sections corresponding to levels of the most characteristic sections have been placed side by side. The entire picture is slightly schematic (Figs. 3 to 12).

The ridge consists of four elements: 1) a nerve issuing from the brain traversing it along its entire length (*N*); 2) a cartilaginous tube which acts as a support for the nerve and gives it its shape; 3) a sensory epithelium
46 (*I*) complicated above the nerve and protruding into the cavity of the organ; 4) a flat epithelium (*E*) externally covering the ridge; it is the same as the vesicular wall.

The nerve coming from the brain is large at its point of arrival on the vesicle (Figs. 3 and 12) and is directly covered by two layers of flat, non-modified epithelium. The sensory epithelium (Figs. 4 and 5) commences development at the point where the cartilage of the ridge starts. The nerve sends fibers to the sensory cells (Figs. 6, 10, 11); it then gradually diminishes and terminates at the other end of the ridge

*Here and in the following pages pertaining to the description of *Bolitaena diaphana* the reference to Plate VII is wrong in the French original. It has been corrected to Plate III—General Editor.

(Figs. 7 and 8) simultaneously with the cartilage and sensory cells. From here onward the two epithelial layers lie in direct contact without nerves or cartilage (Fig. 9).

This is a rough outline of the general arrangement of the organ. A clear idea can also be given of the relationships of various constituent elements which will now be described.

The superficial epithelial layer of the vesicle extends on to the ridge; it is flat and does not exhibit any noteworthy feature. The internal epithelium of the vesicle is ectodermal in origin and simply covers the internal surface of the statocyst; however, on the ridge and spots it becomes the support for the sensory cells. It separates them from each other and for this its structure is considerably modified. Some of its cells are low and short and remain intercalated between the base of the sensory cells like connective tissue cells; some are cylindrical and others are conical and elongated. They are high and, through the intervals between them, the tips of the sensory cells protrude beyond the general level of the epithelium.

The cartilage is a mixture of fibers, isolated polygonal cells and hyaline matrix. It is very poorly developed and has no real rigidity.

The sensory epithelium consists of long, basally broad cells apically terminating in a short, fine point (Pl. III, Figs. 10, 11, 15, *A*) which protrudes into the fluid filling the statocyst. These cells are basally connected with the nerve fibers issuing from a longitudinal nerve cord. They can be clearly distinguished in some histological sections. They are variable in size, larger in the middle of the ridge than at the tips. They are filled with a granular substance and enclose a large nucleus.

The nerve cord lies against the cartilage and at places encloses some cells, probably connective tissue.

The sections of the leaf-like spot show a structure similar to that of the crest; one sees here the same sensory cells embedded in the support tissue (Pl. III, Fig. 14). The cartilage is well-developed.

Otoliths were probably present, but have disappeared during the preservation in formalin; some of the larger cells of the spot probably were designed to secrete these.

This description indicates that the statocysts of this cephalopod, whose habitat is distinctly abyssal, are
47 highly developed. Their large size, transparency and the absence of a solid, opaque cranium helps identify several points of their structure. These characteristics probably also exist in other cephalopods where it is impossible to see them due to their opaque solid structure and reduced proportions.

SUBORDER DECAPODA LEACH, 1818

Division A. MYOPSIDA d'Orbigny, 1845

Family I. SEPIOLINI Steenstrup, 1861

Genus *Heteroteuthis* Gray, 1849

Heteroteuthis dispar (Rüppell) Gray, 1849
(Pl. X, Figs. 1–3)

1845. *Sepiola dispar*, Rüppell (fide Vérany) (**68**), p. 26
1839. *Rossia dispar*, Vérany (**79**), p. 63
1849. *Heteroteuthis dispar* Gray (**22**), p. 90

Expedition of 1912: Stn. 3256, depth 0–1,200 m. August 19, 38°16′40″ N, 28°23′ W. Bourée net rapidly retrieved. Azores.

Only one individual was captured. It is 42 mm long, excluding the rolled tentacles.

To date, no satisfactory illustration of this species has been given. Two beautiful live specimens in perfect condition were collected at Messina in December 1916. Miss de la Roche made magnificent watercolor sketches of these which reproduced in the plate. The large subcutaneous, ventral, luminous organs resembling large silvery plates can be seen here.

Geographic distribution. This species is known in the Mediterranean and the Azores.

Genus *Rossia* Owen, 1834

Subgenus *Franklinia* Norman, 1890

Rossia caroli Joubin, 1902

1902. *Rossia caroli* Joubin (**36**), p. 138
1902. *Rossia caroli*, Joubin (**37**), p. 180
1914. *Rossia caroli*, Joubin (**41**), p. 35

Expedition of 1913: Stn. 3473, depth 1,332 m. September 6, 42°36′30″ N, 63°36′30″ W. Trawl, south of Halifax.

48 One average-size specimen with no trace of hectocotylization was caught.

Geographic distribution. To date, this species has been found in the North Atlantic: the Azores, south of Halifax and in the Mediterranean.

Subgenus *Rossia*

Rossia palpebrosa Owen, 1835

1835. *Rossia palpebrosa* Owen (**60a**), p. 92

Expedition of 1912. Stn. 3340. September 17 to 30. En route to Thamshamn (Norway).

A specimen approximately 9 cm long, including the tentacles and without any trace of hectocotylization was caught in very good condition. This cephalopod conforms well to the type and locality described by the authors. Outside polar waters, it seems to have been encountered only once by Jatta at Naples. I consider this identification highly doubtful.

Geographic distribution. Atlantic, Arctic.

Division B. OEGOPSIDA d'Orbigny, 1845

Family ONYCHOTEUTHIDAE G. Pfeffer, 1900

Genus *Onychoteuthis* Lichtenstein, 1817

Onychoteuthis banksi (Leach) Férussac, 1817

1817. *Loligo banksi* Leach (**49**), p. 141
1839. *Onychoteuthis banksi*, Férussac and d'Orbigny (**60**), p. 330
1896. *Ancistroteuthis lichtensteini* Jatta (**30**), p. 103
1900. *Teleonychoteuthis krohni* Pfeffer (**62**), p. 159

Expedition of 1912: Stn. 3207, surface. August 6, 30°54′ N, 24°11′ W. Near Madeira.

One specimen, 165 mm total length, caught between 9 and 10:45 p.m. in a shrimp net under an electric spotlight. It is almost adult and in very good condition. Stn. 3286, depth 0–1,000 m, August 25, 39°19′ N, 35°404′ [sic; 40′?] W. Bourée net rapidly retrieved. Azores.

One young individual, 90 mm total length including the tentacles. It is in good condition except one eye is missing.

Geographic distribution. This cosmopolitan species is collected from the most diverse localities of the polar and equatorial oceans.

49 **Family II. ENOPLOTEUTHIDAE G. Pfeffer, 1900**

Genus *Abraliopsis* Joubin, 1896

Abraliopsis pfefferi Joubin

1896. *Abraliopsis pfefferi* Joubin (**33**), vol. 5, No. 1
1912. *Abralia* (*Abralia*) *pfefferi*, Joubin (G. Pfeffer) (**65**), p. 156

Expedition of 1912: Stn. 3202, depth 0–3,000 m. August 5, 31°45′ N, 20°17′ W. Bourée net rapidly retrieved. Parts of Madeira. Stn. 3217, depth 0–1,000 m. August 8, 30°45′30″ N, 25°47′ W. Bourée net rapidly retrieved. Madeira region.

The specimen from station 3217 is 57 mm long; it is whole and in very good condition. The luminous organs are aligned as in the typical form and those on the ends of the ventral arms also conform to this type.

The individual of station 3202 is approximately 100 mm in total length including the tentacles; unfortunately one is lost and the other is damaged; however, it has its hooks. The ventral arms have deteriorated somewhat, but one still has terminal luminous organs. The species has been identified on the basis of these two features.

Geographic distribution. Atlantic, Mediterranean.

Abraliopsis sp., juv.

Expedition of 1913: Stn. 3526, depth 0–2,000 m. September 29, 38°30′ N, 34°40′ W. Large Richard net.

This is a very young individual whose total length is barely 1 cm, including the arms. The suckers are not distinct. There are no luminous organs on the ventral side of the skin. A few are seen on the ocular globe.

Genus *Abralia* Gray, 1849

Abralia morrisii Vérany, 1837

1837. *Onychoteuthis morrisii* Vérany (**78**), p. 100
1839. *Enoploteuthis morrisii*, Férussac and d'Orbigny (**60**), p. 338
1849. *Abralia morrisii*, Gray (**22**), p. 50
1900. *Abraliopsis morrisii*, G. Pfeffer (**62**), p. 168
1908. *Abralia morrisii*, G. Pfeffer (**63**), p. 291
1912. *Abralia morrisii*, G. Pfeffer (**65**), p. 168

Expedition of 1911: Stn. 3064, depth 0–2,700 m. July 21, 38°03′30″ N, 4°46′45″ E. Bourée net rapidly retrieved. Mediterranean. Southeast of Minorca. Stn. 3131, depth 0–3,500 m. August 16, 35°09′ N, 21′ W. Bourée net rapidly retrieved. Azores.

50 Expedition of 1912: Stn. 3202, depth 0–3,000 m. August 5, 31°45′ N, 20°17′ W. Near Madeira. Bourée net rapidly retrieved.

The two cephalopods from stations 3064 and 3131 are in very good condition and closely resemble each other. There is scarcely any detail that distinguishes one from the other.

The smaller specimen caught at station 3064 is 60 mm in total length, including the tentacles. Its two ventral arms have typical luminous organs at their ends; each consists of two black pearls. The second specimen from station 3131 is a little larger, 64 mm; its ventral arms also have luminous organs, but it is composed of three black pearls. I do not think this difference alone is sufficient to affiliate this cephalopod to a species of the same genus other than *Abraliopsis pfefferi* Joubin.

I have a few reservations regarding the determination of the small cephalopods from Stn. 3202.

The lot comprises seven small specimens.

These cephalopods very probably belong to this species, but they are not in good enough condition that they can be identified with certainty. Some are missing one or two tentacles, others have none or only one of the luminous black pearls on the ventral arms. The other features agree well with the diagnosis of the species. The number of ventral luminous organs increases with the size of the animal and ultimately constitute rows in the older forms, whereas they are scattered without apparent order in the younger forms. The size of these cephalopods, in mm, from the tip of the arms to the posterior end of the body is as follows: 33, 34, 37, 42, 43, 60 and 62. These measurements are approximate because the arms were curled up and hard and thus difficult to flatten without breaking.

Enoploteuthidae

Genus? Species?

Expedition of 1913: Stn. 3415, depth 0–2,000 m. August 10, 40°15′ N, 56°17′. Bourée net rapidly retrieved. Sigsbee Trench, south of Newfoundland.

This cephalopod is unfortunately in such poor condition that it is impossible to describe it. The form, however, appears to be very interesting. The tentacles and eyes are unfortunately missing. All the arms are truncated and stripped and it is only at the base that some small remaining suckers show that they have brachial hooks. The cone-shaped body has very large fins extending almost to the posterior tip, but do not reach the
51 absolute edge of the mantle. These fins are 35 mm high and 55 mm wide together. Two whitish symmetrical bodies, which are probably photophores, can be seen under the skin at the bottom of the abdominal wall, about 11 mm from the end of the body. I did not find these organs in the mantle cavity or on the rest of the skin or the remnants of the socket. I cannot attribute a name to this animal that defies identification.

Genus *Thelidioteuthis* G. Pfeffer, 1900

Thelidioteuthis alessandrinii (Vérany), 1851

1851. *Loligo alessandrinii* Vérany (**80**), p. 99
1857. *Enoploteuthis polyonyx* Troschel (**77**), p. 67
1880. *Onykia caribaea* Steenstrup (**74**a), p. 96
1882. *Abralia magalops* Verrill (**82**), p. 364
1884. *Enoploteuthis pallida* G. Pfeffer (**61**), p. 18
1889. *Abralia polyonyx*, Carus (**8**), p. 449
1889. *Calliteuthis alessandrinii*, Appellöf (**1**), p. 27
1900. *Thelidioteuthis polyonyx*, G. Pfeffer (**62**), p. 167
1912. *Thelidioteuthis alessandrinii*, G. Pfeffer (**65**), p. 178

Expedition of 1912: Stn. 3190, depth 3,000 m. July 27, 34°27′ N, 14°15′ W. Bourée net rapidly retrieved. In the environs of Madeira.

A specimen in rather poor condition was caught at this station. However, the disposition of the fins which insert along the entire length of the dorsomedian line can be distinguished. The distribution of the suckers, hooks and luminous organs are normal. The specimen is 70 mm in total length.

Geographic distribution. Mediterranean, Atlantic coast of the United States, Antilles, South Atlantic, Canaries, Society Islands and Indian Ocean.

Expedition of 1911: Stn. 3129, surface. August 15, 34°05′ N, 19°37′ W. Fine narrow net; caught at 7:30 p.m. Azores.

The identification of the second specimen is doubtful.

It is a very young cephalopod whose body is 7 or 8 mm long. I affiliate it to this species tentatively since it is in very bad condition, having lost most of its suckers.

The interest in this specimen lies in the large number of varied parasites it bears. This is a very rare occurrence in cephalopods where the presence of parasites is exceptional. I only report their presence and am not responsible for their study. I shall mention only the most important points. A nematode is visible inside the skin of the mantle wall. Two large, soft and rounded tumors are present on the ventral side of the left fin. A small crustacean is lodged in the right orbit. An entire batch of small spherical eggs on a single plane lie against the skin of the ventral side of the head. This description of the parasites is incomplete because some had fallen off during the handling.

52 **Genus *Pyroteuthis* Hoyle, 1884**

Pyroteuthis margaritifera (Rüppell), 1844

1844. *Enoploteuthis margaritifera* Rüppell (**68**)
1900. *Pterigioteuthis margaritifera*, Pfeffer (**62**), p. 166
1902. *Pyroteuthis margaritifera*, Hoyle (**26**), p. 1
1903. *Enoploteuthis margaritifera*, Lo Bianco (**53**), p. 171
1908. *Pyroteuthis margaritifera*, Chun (**12**), p. 87
1912. *Pyroteuthis margaritifera*, Pfeffer (**65**), p. 196

Expedition of 1912: Stn. 3217, depth 0–1,000 m. August 8, 30°45′30″ N, 25°47′ W. Bourée net rapidly retrieved. Close to the Monaco Trench. Stn. 3279, depth 0–3,000 m. August 23, 38°55′ N, 34°07′30″ W. Bourée net rapidly retrieved. Azores. Stn. 3281, depth 0–4,000 m. August 24, 39°25′40″ N, 35°14′30″ W. Bourée net rapidly retrieved. Azores. Stn. 3284, depth 0–1,000 m. August 24, 39°19′ N, 35°24′30″ W. Bourée net rapidly retrieved. Azores.

This small series contains five adult individuals with total lengths varying from 85 to 105 mm including the tentacles. In all specimens, the eyes have come out from their sockets, showing the luminous organs with their red and green iridescent sheen fixed on the ocular globe. The skin is generally intact except for a few rounded shredded places caused by friction. None show any trace of a hectocotylus.

Geographic distribution. Mediterranean, Azores. Amboine (*Challenger*).

Pyroteuthis margaritifera Rüppell, 1844;
var. *aurantiaca* nov. var.
(Pl. X, Figs. 4 and 5)

Expedition of 1911: Stn. 3147, depth 0–550 m. August 26, 38°12′ N, 25°49′ W. Bourée net rapidly retrieved. Near San Miguel, Azores. Stn. 3150, depth 1,740 m. August 27, 38°01′ N, 25°21′ W. Trawl. Near San Miguel, Azores.

Two specimens in very good condition; they definitely belong to the species *Pyroteuthis margaritifera* Rüppell. However, they have some special characteristics which differentiate them from other specimens of the same species that have already been described or illustrated.

The overall color is reddish with a bronzed iridescent background. The dorsum is much darker with large chromatophores scattered, particularly transversely.

The posterior end of the body is more pointed and longer than normal. The yellowish fins are slightly smaller.

The eyes have luminous organs similar to those shown by Chun, but in addition to the seven counted by this author there are two other very small luminous organs that are not normally found. This is an important
53 point, because the arrangement of the luminous organs is absolutely constant and characteristic in this species.

The two specimens are 80 and 85 mm total length including the tentacles.

Pyroteuthis sp., juv.

Expedition of 1911: Stn. 3089, depth 0–4,000 m. July 30, 32°21′30″ N, 12°31′ W. Large Richard net. Between Gibraltar and the Ilheus Selvagens. [North of the Canary Islands].

The cephalopod under discussion is very young and its total size, including the arms, is less than 10 mm. Its specific identification would be quite impossible, if one could not distinguish clearly on the ocular globes the main luminous organs characteristic of *Pyroteuthis margaritifera* Rüppell. The tip of the protruding gladius can also be very clearly seen behind the fins. However, I do not wish to attribute a definite species name to it because I have not been able to see all the luminous organs. Possibly, all these had not developed because of the young age of the specimen. Moreover, I could not examine the individual as thoroughly as required without damaging it as it was too small. I therefore hesitate to attribute a species name to this specimen, in spite of the probabilities.

Family III. PARATEUTHIDAE Joubin, 1919

Genus *Cycloteuthis* Joubin, 1919

Cycloteuthis sirventi Joubin, 1919
(Pl. VII, Figs. 3–6; Pl. VIII, Figs. 1–11)

1919. *Cycloteuthis sirventi* Joubin (**45**[ter]), *Bull. Monaco*, No. 351

Expedition of 1912: Stn. 3213, depth 0–500 m. August 7, 30°45′40″ N, 25°47′ W. Bourée net rapidly retrieved. Night catch. Strait of Madeira.

This is a highly interesting cephalopod. In fact, it shows several characteristics which may even indicate its exclusion from all the families of Oegopsida as they are actually defined. Therefore, no other course is open except to create a new name for it. To be more accurate, I would say that I am re-establishing the family Thaumatolampadidae established by Chun. The creation of this family was not accepted by G. Pfeffer as indicated in his monograph on the Oegopsids of the Plankton Expedition. For reasons of priority, he found it necessary to transform it into a subfamily Lycoteuthinae and incorporate it into the family Onychoteuthidae.

However, the large family Onychoteuthidae is divided into two absolutely unequal sections.

54 1. The subfamily Onychoteuthinae which includes genera and species with hooks; 2. subfamily Lycoteuthinae without hooks.

The second section was established only for the single species named *Thaumatolampas diadema* by Chun and *Lycoteuthis diadema* by Pfeffer for reasons of priority.

For my part, I believe that this cephalopod is far removed from Onychoteuthidae, not only because it has no hooks but also because of other anatomical characteristics, and that there is no adequate reason for keeping it in this family. I believe I must follow Chun, not Pfeffer, and retain the family Lycoteuthinae.

But this family Lycoteuthinae is based until now solely on the features of a single species *Lycoteuthis diadema*. I now bring a new genus to this meager family, with a single species: *Cycloteuthis sirventi*. Some of the features of Lycoteuthinae apply well to it, but there are others that must be modified, and others that must be added. The result is that the definition of the family Thaumatolampadidae of Chun equivalent to Pfeffer's subfamily Lycoteuthinae is no longer acceptable after the introduction of *Cycloteuthis*. It must be therefore modified, but then my definition will differ from that of preceding authors while bearing the same name, thus causing confusion. The name of the family has to be changed so that Joubin's diagnosis is not confused with either Chun's or Pfeffer's which no longer apply well to the two constituent genera.

I have given it the name Parateuthidae Joubin which includes two genera, 1. *Lycoteuthis* G. Pfeffer (=*Thaumatolampas* Chun) and *Cycloteuthis* Joubin.

The problem now arises of the placement of this family. The fins in this family definitely have similarities with those in the Onychoteuthidae. However, the arms and tentacles do not have hooks. It thus lacks a fundamental characteristic as well as other less important features that would affiliate it to this family as Pfeffer had proposed. For the same reason, it cannot be considered as affiliated to Enoploteuthidae. The mantle is not fused with the funnel, a feature that distinguishes the form from the Cranchiidae, Galiteuthidae, etc. It does not have affiliations with the Chiroteuthidae and still less with the Ommatostrephidae. There remains Histioteuthidae; it differs from this family by the fins, the absence of surface luminous organs except those of the eyes, but resembles it in the arms, tentacles, suckers and some other details. I believe that this family lies rather naturally between the Histioteuthidae on one hand and Onychoteuthidae and Enoploteuthidae on the other, as can be judged from the following description.

I have unfortunately only one badly damaged specimen of this cephalopod. I could conduct only an incomplete study, as I did not wish to completely destroy the specimen of the type genus and species to be preserved in the Museum at Monaco.

55 Measurements (in mm)

Total length including tentacles ..	150
Length of 1st arm (dorsal) ..	35
Length of 2nd arm ..	50
Length of 3rd arm ..	49
Length of 4th arm (ventral) ..	41
Length of tentacle ..	88
Length of tentacular club ..	17
Length of body including the posterior tip ..	49
Length of fin ..	30
Width of the two fins ..	48

A superficial examination of the specimen indicates that it belongs in the Onychoteuthidae, or in *Enoploteuthis*, or in *Abralia* or *Pyroteuthis*. The shape of the body and the fins, the proportion of the arms, tentacles and suckers impart this initial thought. However, one looks in vain for hooks on the brachial suckers which have minute teeth so small that they cannot be detected at first. They are sharper but very small on the tentacle. The general shape, color and firmness of the tentacle affiliate it to Histioteuthidae.

When alive the color of the body is reddish with lighter arms and fins.

The fins are large, muscular, high, rhomboidal and give the impression of a strong swimmer with solid, thick and strong muscles contrasting with the softness and usual infiltration of the tissues of the Histioteuthidae which they resemble in other points of their organization.

The body is cone-shaped and terminates posteriorly in a sharp point extending approximately 1 cm beyond the fins. The cone gradually broadens upward and its margin rises higher dorsally than ventrally. The dorsal mantle border is only approximately 1 cm above the upper angle of the fin. This indicates that the fin is very large and its median depression gives it a strong resemblance to that of *Octopodoteuthis sicula* Rüppell.

The two fins are symmetrical and meet on the dorsomedian line, except in the depression mentioned above. This character removes *Cycloteuthis* from Histioteuthidae in which the fins are separated by a triangular space. The fins are noticeably rhomboidal in shape with two rounded, extreme angles. Bundles of parallel muscles issue perpendicularly to the gladius to reach the edge of the organ and can be clearly distinguished on its ventral side.

56 The body and fins still have shreds of skin on which no trace of luminous organs could be detected. This is an important difference from Histioteuthidae. The chromatophores are rather numerous, small and uniform but do not seem to be placed in several layers.

The projecting end of the body and the few millimeters preceding it have a gelatinous consistency and are semitransparent through which the terminal point of the gladius can be seen.

The funnel is highly developed and strongly projects from the border of the mantle slit; its opening is located almost at the level of the middle of the eyes. It is slit from the lower side, exposing the complicated organ it encloses. Two oblique, approximately 4 mm long funnel locking cartilages can be seen on the margins of this slit. Their edge is formed of a similar, strongly protruding ridge that encircles an oval-shaped groove with a flat base; it is surrounded by a line of long, closely-packed chromatophores. A small internal cartilaginous protuberance simulates a barely visible rudimentary tragus. The corresponding mantle ridge is slightly protuberant, lightly arched and a little higher in the middle than at the ends.

If the funnel is split along the sagittal line, a highly developed and simultaneously very complicated internal apparatus can be seen. A semicircular valve is situated at the top below the opening in the form of a transverse slit of the funnel. The valve is very thin with a convex upper face and is almost the same diameter as the opening of the funnel. Below this lies the main part of the *funnel organ* which occupies almost the entire length of the funnel from the valve. It is difficult to describe because of its complexity. It can be represented as consisting of a pedestal of thickened skin which is triangular, almost like the letter A. A membranous, opaque protruding crest is on the two oblique branches. They do not meet at the apex of the angle where a third very tiny, triangular, vertical branch is situated. The two lateral crests are separated by a ridge of this small unpaired crest. The two large lateral crests rise basally and terminate when buried in the general base. The edge of these two crests is flexuous, wavy and strongly protuberant in its middle part.

Two other pads are located on each side of this central patch. These are semicircular, similar to the valve, but much larger and obliquely inserted on the funnel wall. They are opaque and have the same consistency as the crests of the central organ. When the funnel is intact they are positioned above the branches of the A; their rectilinear bases almost touch each other. The tubular cavity of the funnel is almost completely obstructed by these various parts of the funnel organ or valves that forces the water traversing the organ out into several jets.

The cephalic vein runs between the two branches of the A and then descends vertically along the length of the visceral mass immediately under the skin to the level of the rectum.

57 The large size of the funnel, the development of the valve, the crests and pedestal that form the funnel organ correlate with the considerable development of the fin and indicate that this cephalopod is a powerful swimmer.

The head of the animal is unfortunately in very poor condition; one of the eyes is entirely missing and the other is badly damaged. However, it can be seen that it was very large. The head, compressed between the two vast sockets, is very narrow. Its tissues, like those of the base of the arms and periphery of the funnel, are semitransparent; it still has its skin but the luminous organs are totally absent. However, a circle of some 20 of them is present inside the skin of the periphery of the eyelid. They are small, yellow, slightly golden and iridescent. The skin cover for the eye is in shreds and thus no details can be given. Luminous organs do not seem to have been present even on the ocular globe.

The buccal bulb is interesting to study. It possesses a distinguishing feature that permits the separation of the genus *Cycloteuthis* from the genus *Thaumatolampas*. The latter has eight pillars and the former has only seven.

The buccal bulb is hexagonal and white with no pigmentation or chromatophores. It is attached to the arms in the following manner. Between the two dorsal arms, there is a single large rib or pillar, the base of which is bifurcate and continues by a small thin lamella on the dorsal edge of these two arms. A large rib on each side issues from the bulb and is attached by a thin membrane to the dorsal edge of arm No. 2. The third pair of arms has the same arrangement except that the thin lamella issuing from the corresponding rib is attached to the ventral edge of the arms. Finally, between the ventral arms there are two large ribs very close together, in fact so close that at first they seem to be one. They are prolonged by two long transparent triangular lamellae which join the ventral edge of the fourth pair of arms. These different lamellae issuing from the ribs are inserted on the edge of the cupuliferous area of the corresponding arm, dorsally for the first two arms and ventrally for the last two.

The ribs are terminated by a slightly bifid tip above the buccal opening. The membrane forming the most external lip of the bulb is between these ribs. It is thin and wavy along the edge and seems to be longitudinally folded. This labial membrane extends along the length of the ribs, joins one to the other and is free on its border like the fabric of an umbrella. The solid wall of the muscular bulb can be seen below.

The protruding, black tip of the lower beak can be seen through the opening of the mouth. It emerges from a papillose lip in the form of a circular pad. Finally, the papillae dispersed within the buccal cavity can be detected, hanging down from the folded skin between the ribs.

58 A white olfactory tubercle, approximately 2 mm long, is under the head at the optic swelling. Its matted appearance contrasts the transparency of the skin. Two wavy nerve fibers which emerge from the deep mass of the head penetrate here.

The nuchal cartilage is well-developed; it is transparent and on the head presents a median ridge bordered by two vertical crests, then two slightly wavy surfaces. A similar but inverse arrangement is observed on the mantle. The entire apparatus is slightly larger at its nuchal tip than it is below and terminates in a slightly wavy curved line. On the whole, the adhesive apparatus is well-developed and is large in proportion to the rather small size of the animal.

The arms are fleshy and thick in relation to their length. Only their tips are slender. Their section is rounded except on the buccal or inner surface which is flattened. This surface is bordered on each side by a slight crest one of which, either dorsal or ventral, continues with the membranous prolongation of the corresponding buccal bulb. The cupuliferous zone, thus defined, is especially well marked on the dorsal arms; the crest delimiting it is more protuberant than in the other arms and the base of the sucker here seems to form small protruding tubercles that festoon it. Moreover, it is observed that brown and reddish-brown chromatophores are very abundant in this zone, whereas they are very sparsely scattered on the dorsal and lateral surfaces of each arm. Several overlapping layers can be seen in this region since the skin is transparent. Each sucker is supported by a small nipple at the tip of which is placed a short, squat peduncle. These nipples have numerous chromatophores which are absent on the sucker and its peduncle.

The arms are soft, flaccid and almost colorless at their tips. There is no web between them. I have found no trace of luminous organs on their surface, base or tip. The suckers are regularly arranged in two alternating rows, but this arrangement is only noticeable on the lower third of the arms; frequently they are in a single row. Those in the middle of the third arm are slightly larger than the corresponding suckers of the other arms.

The suckers of the arms are spherical, white and supported on a cylindrical peduncle which is much shorter than the slender body of the organ. The part where the peduncle is inserted is slightly flattened. The opening, surrounded by a yellow and brown horny ring, occupies approximately three-fourths of its height and it is slightly oblique. The height of these suckers is .25 to 2 mm; those in the middle of the third arm reach 2.5 mm.

The funnel-shaped horny ring defined by an outer layer and a dark brown base is unequal; its lower edge is broader than its upper edge. This edge has small ornaments in the form of slight polygonal protuberances; they are in four rows in the part of the circle nearest to the peduncle. There are only three rows on the sides,
59 then two, and finally only one in the upper fourth of the opening. The first row of these small tubercles closest to the deep opening consists of larger and darker tubercles; the others are gradually less dark and clear as they approach the free edge.

The horny ring also has teeth on the third that is farthest from the peduncle; they are extremely low, flat, without denticles and, at first, it is rather difficult to see them. The two middle teeth are more distinct, the next two on the right and left are still visible but the fourth is so low and so indistinct that it seems to be a tiny brown crest on the horny ring. There are eight teeth in all.

The tentacles are strong, fleshy, much more consistent and solid than the arms. They are pigmented brick-red with fine reddish-brown streaks. Their inner surface is slightly flattened and a barely protuberant, very slight keel can be seen along the length of the outer surface; this keel is slightly more accentuated at the club where it is completely transformed. Suckers are absent on the tentacle and they begin only at the lower limit of the club.

The club is generally lanceolate in shape. The cupuliferous side starts with five small suckers with three knobs on one of the tentacles and three suckers with five knobs on the other. The sucker-knobs have a small, horny, brown ring without teeth, but small protuberances are present on the funnel-like plate. The knobs have no chromatophores or pigment; they stand out in white against the brick-red skin surrounding them; they have the appearance of a small hemispherical cap. These knobs and suckers are sessile. They are present near the extremity of the flat surface of the tentacle, immediately before the beginning of the club itself. This part is highly complicated. Its oral surface which has suckers and is continuous with the flattened surface of the tentacles, is in two parts; the lowest part is limited on each side by a marginal membrane consisting of skin with chromatophores supported at intervals by a whitish stalk; there are seven or eight of these stalks on each side. This is the part of the club that supports the large suckers in four rows. This part forms about two-thirds of the club. The distal third is occupied by smaller suckers in four rows but not demarcated by the membrane with stalks. Moreover, this region is vertically flattened and the small suckers appear to be placed on one side but not on a line with the larger suckers. I might say that it is the dorsal keel of the club that is extended to carry the small suckers, instead of being a continuation of the oral surface with large suckers. Finally, at the very end of this club, the four rows of very small suckers separate into two rows of two and reunite a little further on to encircle a small round surface without suckers. Something similar is seen in *Thaumatolampas diadema* Chun.

60 If the club is turned over, it can be seen that its dorsal surface has a very low crest in the first half of the region with large suckers. The crest then becomes more accentuated and finally very large in the region with small suckers. It becomes so large that it alone seems to continue on the upper third of the club and holds the four rows of small suckers on its crest. The club is obviously very complex.

The suckers are no less so. They all have a rather long peduncle, as long or even longer than the body of the sucker itself. But the peduncle is so implanted on the club that the sucker itself is much higher than the point of insertion of its peduncle. All the suckers are thus directed toward the tip of the club and the peduncles toward its base.

The diameters of the large suckers are 1 to 2 mm, and the small ones are 1/8 to 1/3 mm. Their horny ring is very large in relation to the fleshy part. It is brown on a background of yellowish-orange. A semicircle of teeth can be observed; these are pointed but low and very dark, almost black. The middle teeth are well developed but they soon diminish and disappear in the lower half. The largest tooth is exactly opposite a slit in the ring. The teeth extend by a brown trail which loses itself inside the sucker on its chitinous body. The edge of the funnel is very complicated; it is striated with innumerable extremely fine small ribs which issue from the internal horny ring to reach the free edge supported itself by a slightly darker, thin circle. These small protuberant, brown ribs cut across the yellow chitinous membrane which supports them. Almost all the ribs are simple, a few are bifurcated. Within the entrance of the funnel below the teeth, there is a black, chitinous flexuous line whose outward protuberances correspond to the intervals between the teeth. This ornamentation is very interesting; it is repeated on all the suckers of the palette.

The radula (Pl. VIII, Fig. 11) is small, 1.5 mm wide. The large spine of the marginal tooth is the most noticeable feature here. The middle tooth has three points, the center one much larger; the middle teeth each have an intermediate point between the median and marginal.

There are two types of luminous organs: some are included in the border of the eyelid, as mentioned earlier in the text; the other, of which there is only one, is visceral. The latter luminous organ will be discussed first.

The visceral photophore is large, unique and oval-shaped; it lays on the visceral mass between the two gills in a loop of the rectum, on the mass of the liver and ink sac which constitute a black screen for
61 it. It is covered by a transparent, convex, oval-shaped cornea embedded in the white integument which reveals the upper part of the ink reservoir. The yellowish mass of the photogenous tissue is visible through the cornea; this mass is oval-shaped and bordered by a black ring which probably is the ink sac, visible because of the transparency. I did not take sections of this organ as I did not wish to destroy it in this unique specimen. We will have to wait until an additional specimen is obtained before we can ascertain its structure. It is noteworthy that a visceral photogenous apparatus consisting of a single luminous source has not been described to date in any other oegopsid cephalopod; there are always several of them, some paired and symmetrical, others on the median line. This one is relatively large since its vertical diameter exceeds 4 mm and alone it replaces all the other very much smaller photogenous organs distributed on the visceral complex observed in similar cephalopods, particularly *Thaumatolampas diadema* where Chun has described seven. This arrangement is thus completely characteristic of this species and is morphologically very interesting.

The photogenous organs of the eyelid (Pl. VIII, Fig. 10), as mentioned earlier in the text, form a continuous circle of about 20 embedded in the transparent and semicartilaginous margin of this membrane. They are small, unequal in size and orangish-yellow with golden reflections.

The gladius is 55 mm long (Pl. VIII, Figs. 8 and 9). It presents a slight dilation in its anterior third, the middle third is cone-shaped, much more prominent than the first. Finally, it terminates in a very sharp curved point. It also terminates in a point at the top, but this is simply the end of the flat rachis. It is broader where it corresponds to the upper dilatation than in its other part. A keel is present for the entire gladius which, as a whole, curves slightly inward toward the visceral mass. The two flaps of the large cone are juxtaposed but not fused. They fuse only at the level of the caudal tip to form the tiny cul-de-sac, following which the full terminal point commences.

The dilated part of the gladius corresponds to the fin, which is highly developed. The slight contraction between the cone and upper dilation is at the depression which separates the fins at the top.

The gladius differs noticeably from that of *Thaumatolampas* which has no pointed tip but a rounded bud. It also does not have the cone of the lower dilation and its terminal upper part is rounded and not pointed. This entity is much weaker than in *Cycloteuthis*, that which corresponds to the greater development of the fin in the second genus.

Family HISTIOTEUTHIDAE Verrill, 1881

This family, which a few years ago seemed to be so clearly delimited and easily defined, has become very complicated. The discovery of new forms, mainly young individuals considered separate species,
62 the application of the rules of nomenclature, the subdivisions and splittings proposed by recent authors have created what now seems to be an apparently inextricable confusion.

In 1910, Chun attempted to review historically the sequential complications of the family. In 1912, Pfeffer made a similar attempt. He presented a completely different pattern and his conclusions, though highly complicated, seemed to provide a basis for new divisions. However, more recently S. Berry discovered some species that could not be assigned in the proposed framework; moreover, he considered several of Chun's species as young forms. Therefore, the best step would be not to insist too rigidly on even provisional classifications; experience has shown that they are far too ephemeral.

We shall present the seemingly most valid suggestions of Pfeffer, Chun and Berry.

If we consider the characters proposed by Verrill, Chun and G. Pfeffer for the family Histioteuthidae we can form two sections based on the arrangement of the luminous organs. The first includes the three genera and subgenera *Stigmatoteuthis*, *Calliteuthis* and *Histioteuthis*. This appears very simple but in reality it is much less clear.

First, the genus *Meleagroteuthis* seems to be easily characterized, not only by its luminous organs but also by the serrated cartilage ornamenting the keel of almost all its arms and the dorsum; no other cephalopod has a similar protection. But now S. Berry has discovered a *Meleagroteuthis* without this feature. More recently he described another cephalopod which is a *Meleagroteuthis* in some respects, particularly its rudimentary cartilaginous plates, but which in other respects is definitely more closely affiliated to *Calliteuthis*, where he finally places it. Here, therefore, the once so clear classification is on the way to becoming confused.

Matters become increasingly complicated if the various known species of the three genera and subgenera are to be incorporated in their sections. Thus, the forms described by earlier authors under the single species of *Calliteuthis reversa* are distributed by Pfeffer into four species and two genera.

Conversely, he unites under the single name of *Histioteuthis bonelliana* (Férussac) the species that he had earlier placed under the names of *H. rüppelli* Vérany, *H. collinsii* Gray and *Histiopsis atlantica* Hoyle.

It is also noteworthy that S. Berry (1918) expressed the opinion that *Calliteuthis hoylei* and *C. chuni* of Goodrich and of Pfeffer are young individuals of other species not yet identified in the adult state.

The uncertainty that prevails is fairly evident. Reference should be made to the works of the aforementioned authors if one desires to follow the history of these changes; it is pointless to review it again.

It is interesting to note that in a considerable number of individuals, there is an asymmetry of the body,
63 mainly of the head in which the eyes and neighboring organs are much larger and more developed than in others. New examples of this will be found later in the text. At present, no explanation can be given for this phenomenon.

However, a provisional solution must be arrived at to study the material of family Histioteuthidae, caught during the voyages of the Prince of Monaco. I give what I have arrived at without thinking that it represents a final arrangement.

Chun accepts only two genera for Histioteuthidae: *Calliteuthis* and *Histioteuthis*. Like Pfeffer, I think the inclusion of a third genus, *Meleagroteuthis* is necessary.

A new difficulty will arise in the distribution of species placed under the genus *Calliteuthis*. We shall first deal with *Meleagroteuthis* before looking for a solution.

Genus *Meleagroteuthis* G. Pfeffer, 1900

Meleagroteuthis hoylei G. Pfeffer, 1900
(Pl. XI, Figs. 1, 2; Pl. XII, Figs. 1–5)

1900. *Meleagroteuthis hoylei* G. Pfeffer (**62**), p. 170
1908. *Meleagroteuthis hoylei*, G. Pfeffer (**63**), p. 292
1912. *Meleagroteuthis hoylei*, G. Pfeffer (**65**), p. 291
1910. *Calliteuthis meleagroteuthis* Chun (**14**)*, p. 170
1912. *Meleagroteuthis hoylei*, G. Pfeffer-Berry (**4**), p. 305

Expedition of 1912: Stn. 3285, depth 0–500 m. August 24, 39°23′ N, 35°18′30″ W. Bourée net rapidly retrieved. Azores, west of Flores Island.

This beautiful species is so interesting because of its unique characteristics that have been described in detail by G. Pfeffer. It has been again discovered by Berry (1910–1912) who published various supplements. Finally in 1905, I gave a description of the luminous organs of the specimens caught during the Dutch *Siboga* expedition.

Geographic distribution. The localities where specimens of this species have been discovered to date are Gulf of Fonseca, west coast of Central America (Pfeffer); Monterey Bay and in the open sea of Santa Barbara Island (Berry); Paternoster Islands (Kepulauan Tingah), Dutch Indes (Joubin), and this specimen from the Azores.

Chun, in his monograph on the Oegopsida of the *Valdivia*, reported a juvenile of this cephalopod but did not describe it. However, he considered it necessary to suppress the young *Meleagroteuthis* and consider it as a subgenus of *Calliteuthis*. Since he gave no justification for this change I do not accept it as justified and retain the name given by Pfeffer, the more so since Chun, contrary to the rules of nomenclature, suppresses
64 the name of the species given by Pfeffer without any explanation and transforms the generic name (*Meleagroteuthis*) to the species name of the same animal. With the adoption of such procedures there is no nomenclature possible.

An interesting fact is mentioned in Berry's memoir. One of the most striking features of this cephalopod is the presence of hard, serrated points implanted in a solid cartilage on the arms and the dorsomedian line of the body. Pfeffer and I have both noted their presence. Their description will be given later in the text. Berry could not detect any similar feature in his specimens. His words are quoted here, "More difficult to understand is the fact that I have been unable to identify any structures corresponding to the rows of tubercles (Höckerreihe)** described by Pfeffer with much stress as occurring on the arms and the dorsal surface of the mantle along the median line; but this, too, may turn out to be a juvenile character."

I cannot accept Berry's explanation. The presence of these characteristic organs in my specimen, which is 195 mm long, and their absence in Berry's individuals, 203 and 168 mm long, shows that this is not a question of age. Perhaps an explanation must be sought in relation to sex, but these specimens have not been adequately studied from this point of view. It remains to be ascertained whether Berry's specimens have been correctly affiliated to this species. All the characteristics cited and the illustrations are in good agreement with the descriptions by the authors, but the absence of serrated cartilages is so disconcerting that I have reservations regarding Berry's identification of his specimens with this species. We will have to await the capture of more specimens to obtain a solution to this problem.

I have also found some differences of detail between the luminous organs of the specimens described by Berry and those of the *Siboga* specimens whose description I published in 1905. I therefore think I must complete the study of these organs and the serrated cartilages in the specimen from the Azores. It is first

*13 in original—Scientific Editor.
**Misspelled in French original (Hoekerreihe)—Scientific Editor.

observed that the photophores are very closely set on the ventral side of the mantle. They are increasingly more widely spaced toward the dorsum and disappear around the notched dorsal crest. I counted 40 of them on the margin of the mantle opening.

At the base of the ventral arms 12 are arranged longitudinally, a little higher up there are ten, and then six toward the middle of the arm. I could not make out those at the end of the arms where the skin has to a great extent disappeared. There are 25 organs between the two eyes, more closely arranged around the left eye than around the right eye due to the characteristic cephalic asymmetry of this cephalopod. Luminous organs are absent on the funnel, fins and tentacles.

The exterior of these luminous organs has two different aspects. Their lower light-producing part consists
65 of a thimble-shaped black nodule with an indented edge applied against the skin. In some of the organs, this part is completely black, in others, a rounded white spot is present in the middle of the free black face of the nodule. The first type is visible near the median ventral line and the second mainly on the sides. This second arrangement is also visible in the specimen described by S. Berry.

The chromatophores are arranged in the skin around the luminous organs almost in the same manner as described for the specimen of the *Siboga* expedition. It may be noted that there are no yellow chromatophores; I found only red ones of various hues and sizes. However, there are always one or two very much brighter chromatophores above the lens and oval mirror. The latter remains brilliant with a silvery sheen, and some have a pearly yellow appearance. They are less clear and dimmer in the organs of the arms.

The measurements (mm) of this cephalopod specimen are given below:

Total length including tentacles	..	195
Total length without tentacles	..	155
Dorsal length of body	..	51
Ventral length of body	..	48
Length of head	..	28
Width between the eyes	..	25
Length of first arm (dorsal)	..	68
Length of second arm	..	70
Length of third arm	..	72
Length of fourth arm (ventral)	..	73
Length of tentacle	..	115
Length of tentacular club	..	16
Length of funnel	..	15
Length of fin	..	23
Width of fin	..	32

The measurements, to a few millmeters for the arms, are not absolutely accurate as the specimen was not in good condition.

I think the rate at which the net was hauled up was responsible for damaging the specimen. A considerable part of the skin was shredded; many of the embedded luminous organs can be seen in the shreds; they are well-preserved on the mantle. The tips of the arms and suckers have suffered greatly. This chaffing of the skin has revealed the serrated cartilages covering the arms and dorsal ridge of the mantle and are illustrated in detail. The eyes are completely missing, adversely affecting measurements of the head.

Pfeffer has reported a very special feature of the head in the specimen of this species he studied; it is
66 very asymmetrical, one side is very much more developed than the other. S. Berry also found this characteristic and his comments are quoted here ''Head enormous, rounded, strongly asymmetrical, deflected to the right so that its longitudinal axis shows a divergence of some 30° from that of the body; this state apparently

brought about chiefly through the extraordinary development of the left eye which includes nearly one-half of the total bulk of the head, the diameter of its lid opening more than twice that of the right.''

A similar asymmetry is found in the specimen from the Azores; it is distinct (Pl. XII, Figs. 3 and 4) but less striking because the sockets are empty as the eyes have been ejected. The left socket is larger than the right and the portion of the head below this eye is lower than the corresponding portion of the right by at least half a centimeter. The axis passing through the eyes, which should be horizontal, inclines strongly toward the left. The contiguous border of 21 luminous organs surrounding the smaller, right eye is another interesting and noteworthy feature. The left eye does not have this border and only a few very small, atrophied, luminous organs are spaced at variable distances from the edge of the eyelid. The organs are sparse around the left eye but abundant around the right.

A row of serrated tubercles on the three pairs of dorsal arms and on the dorsomedian line of the mantle is one of the most striking features of this cephalopod. They have been reported by Pfeffer, but S. Berry has reported their absence. In the larger specimen under discussion here, the deterioration of the skin and its partial removal has exposed the cartilaginous plates, enabling detailed observations (Pl. XII, Fig. 1). Each consists of a hard, thick, dermal cartilaginous section rising into saw-like teeth along the median line. The teeth are almost all similar. However, a smaller tooth is occasionally intercalated between two larger ones, and sometimes two or three teeth fuse to form a dentate plate. I observed 25 teeth and a few very small supplementary teeth on the first dorsal arm, 18 on the second, 12 on the third and none on the ventral arm. There are 18 teeth on the dorsal crest of the mantle.

Each tooth is laterally flattened like the teeth of a saw; they are very accurately aligned without deviation. They are apically yellowish, the hue is paler toward their base. This base is encircled by a very faint groove looped around the series of teeth. The tip is soft and is perpendicularly flattened from top to bottom due to the lateral flattening of the entire tooth. This tip is very hard but does not form a hook.

The mantle has a slightly different dentation. The teeth are embedded in the cartilaginous band with only their tips protruding. The teeth of the arms have very faint secondary denticulations of their cutting edge.
67 Here, however, they are smooth, more irregular and slightly covered. The difference in appearance is noticeable although the dimensions are the same. The teeth provide an immediately noticeable characteristic differentiating the denticulate cartilage of the arms from that of the mantle. That characteristic is that the brachial teeth are independent of each other and not basally contiguous, whereas the mantle cartilage has more the appearance of a continuous band of cartilage from which the tips of the teeth emerge.

The peribuccal region is very complicated. A small, slightly developed web is between the dorsal arms. An even smaller one is found between arms one and two. It is missing between the other arms.

The buccal membrane between the two dorsal arms has one tip exactly on the median line where it bifurcates. A median tip is also present between arms one and two on each side. However, the four angular ventral tips are continuous with the ridge that carries the ventral row of these four arms (arms three and four encompass the tentacles). This makes seven tips in all.

Genus *Calliteuthis* Verrill, 1880

As stated earlier in the text, considerable difficulties arise in the interpretation of the genus *Calliteuthis*. It can be accepted *sensu lato* as Chun does or *sensu stricto* following Pfeffer or even in a third form that I shall term *sensu medio*. This one is derived from Chun after removing the genus *Meleagroteuthis*.

If we concur with Pfeffer's views, there is only a single species in the genus *Calliteuthis* Verrill 1880: *C. meneghinii* Vérany, 1851. For reasons of priority, Pfeffer considered that this name had to be substituted for *C. reversa* Verrill, accepted by almost all authors since Verrill. It is the former *Loligo meneghinii* Verany, 1851. There is no doubt that this species does not belong in the genus *Loligo*, but it is not undoubtedly proven that *Loligo meneghinii* Verany 1851 = *Calliteuthis reversa* Verrill 1880. We will admit the latter synonymy so as not to complicate the discussion any further.

If we agree with the views of Chun, then the genus *Calliteuthis* Verrill will have to be maintained, but we would have to also establish the subgenus *Stigmatoteuthis* Pfeffer that comprises the two following species: *Calliteuthis* (*Stigmatoteuthis*) *hoylei* Goodrich, and *C.* (*St.*) *ocellata* Owen. We have just stated that *Meleagroteuthis*, considered by Chun as its second subgenus, will be removed from the genus *Calliteuthis* Verrill.

Pfeffer described seven species of his genus *Stigmatoteuthis*. Not having made close observations on these other than knowing them from the memoirs of the authors, I do not have a well-based personal opinion on their validity. However, I can state that I am not the only one to have doubts regarding their present position and do not consider it as definitive. For example, *Histiopsis hoylei* Goodrich is considered *Stigmatoteuthis hoylei* (Goodrich) by Pfeffer and *Calliteuthis* (*Stigmatoteuthis*) *hoylei* (Goodrich) by Chun. Finally, S. Berry suggests that it is a young form of an unidentified adult species. There are other complications to consider regarding this species. In fact, Pfeffer created a new species named *Stigmatoteuthis chuni* and considered it synonymous with the *Calliteuthis hoylei* by Chun; consequently this latter could not be the same animal the one Goodrich named *Histiopsis hoylei* and which Chun would have named *Calliteuthis hoylei.* I cannot explain these contradictions from the author that are found on pages 281 and 285 of Pfeffer's monograph (**62**).

Genus *Calliteuthis* Verrill, 1880

Calliteuthis meneghinii (Vérany), 1851

1851. *Loligo meneghinii* Vérany (**80**), p. 98
1880. *Calliteuthis reversa* Verrill (**81**), p. 393
1900. *Calliteuthis reversa*, Joubin (**35**), p. 72
1900. *Calliteuthis reversa*, G. Pfeffer (**62**), p. 170
1907. *Histioteuthis bonelliana* Massy (**55**), p. 381
1909. *Calliteuthis reversa*, Russel (**69**), p. 446
1910. *Calliteuthis reversa*, Chun (**14**), p. 174
1912. *Calliteuthis meneghinii* G. Pfeffer (**65**), p. 250

Expedition of 1912: Stn. 3217, depth 0–1,000 m. August 8, 30° 45′ 30″ N, 25°47′ W. Bourée net rapidly retrieved. Between Madeira and the Azores. Stn. 3281, depth 0–4,000 m. August 24, 39°25′40″ N, 35°14′30″ W. Bourée net rapidly retrieved. Azores, west of Flores Island. Stn. 3285, depth 0–500 m. August 24, 39°23′ N, 35°18′30″ W. Bourée net rapidly retrieved. Same locality as the preceding station. Stn. 3312, depth 0–3,500 m. September 1, 45°02′ N, 13°05′ W. Bourée net rapidly retrieved. In the sea off Cape Finisterre.

Expedition of 1913: Stn. 3473, depth 1,332 m. September 6, 42°36′30″ N, 63°36′30″ W. Trawl. South of Nova Scotia.

Expedition of 1914: Stn. 3591, depth 0–3,500 m. July 26, 42°18′ N, 12°08′ W. Bourée net, closing type. In the sea off the coast of Portugal.

The lot of specimens caught at station 3591 shows a few interesting peculiarities. First, it represents the only cephalopod material caught during the 1914 cruise which lasted for only a few days as it was interrupted by the beginning of the war. Next, it is noteworthy that the catch consists of 11 individuals of the same species and of similar appearance; all have an equal degree of asymmetric head deformation. Initially, when only a few specimens were collected and studied it might have been considered that this deformation was due to the effects of preservation. It is now certain that it is actually a normal and general modification affecting
69 the entire left side of the head. This does not make the feature any less inexplicable and, interestingly enough, it is observed in all the genera of the family Histioteuthidae. It would be worthwhile to examine the collections

of specimens studied by earlier authors to verify whether this asymmetry has passed unnoticed.

All these specimens of *Calliteuthis meneghinii* from station 3591 are 80 to 135 mm long including the tentacles. They are of different sizes but it may be remarked that differences in the state of extension or contraction of the tentacles may have possibly resulted in significant differences in their total length, whereas the body itself varies much less. These cephalopods can be considered juveniles almost at the adult stage. Perhaps they were members of a large shoal of individuals of the same age, since 11 very slightly differing individuals were caught in the same haul.

The right eye is noticeably smaller than the left in all the individuals; the difference is apparently one-third the diameter. The right eye, which is normal, has a circular eyelid bordered by 10 to 20 luminous organs of almost uniform size forming a kind of circlet of golden pearls around the round orifice. The luminous organs are numerous and large on all this half of the head. Conversely, the eyelid of the left eye is irregularly shaped and depressed at the top; the opening is larger than that of the right side. Here five or six luminous organs form an arched line over scarcely one-fourth of their extent; they are on top of the eye slightly away from the edge of the lid. There are only a few small and irregularly dispersed organs on the side of the head around the eyelid, but none on the inferior border. Luminous organs and chromatophores are very sparse on the skin of this entire region of the head.

On the third arm particularly to the right, a small semicircular membrane forms a protruding crest toward the middle of its outer ridge; it is only visible in this part and seems to be equally indistinct in all individuals.

The horny ring of the largest suckers of the club does not seem to be very firm. It bends so easily that in some individuals it is flattened and biscuit-shaped. The arrangement of the suckers on the club in all these individuals corresponds very well to Pfeffer's representation of it (Figs. 13 and 14, Pl. 22). The small suckers and adhesive knobs are similarly arranged here. No teeth were detected on the horny ring of the suckers of the arms and tentacles. Therefore, there is no reason to attribute these cephalopods to any species of the genus *Stigmatoteuthis* of Pfeffer.

The specimens from the other stations are less interesting.

The specimen from station 3312 is only a fragment consisting of the mantle and the lower half of the head; it is about 65 mm long. This must be the largest specimen; the arrangement of the luminous organs is so similar to the arrangement in the other individuals that I think the species identification is accurate.

70 The specimen from station 3281 is mutilated and has a total length of 85 mm. The arrangement of the luminous organs in three rows on the ventral arms and two on the lateral arms corresponds well to *Calliteuthis reversa* of Chun. However, extremely small denticulations are detected with difficulty on the outer edge of the horny ring of the brachial suckers. These are probably the horny accessory organs mentioned by Pfeffer but which cannot be considered teeth comparable to those of *Stigmatoteuthis*. As suggested by Berry for a single species, it is highly probable that these minimal denticulations vary from individual to individual and are thus without specific value.

The usual asymmetry is seen in the specimen from station 3285. This animal is 125 mm long and the asymmetry affects not only the left part of the head but also extends to the corresponding arms where there are far fewer luminous organs than on the symmetrical right arms. Among the rows of large ventral luminous organs on the mantle, small developing ones appear to be alternating.

A few luminous organs are on the specimen from station 3217, which is 73 mm long without the tentacles, which are missing. This shows an interesting stage in the sequence of the appearance of these organs. A continuous marginal series along the ventral mantle margin makes a three-fourths turn along the mantle slit; below this there is a row of some ten organs almost as large, then another alternating row contains fewer and smaller organs; five successively shorter rows with increasingly smaller elements are between the mantle margin and the posterior tip. This arrangement is interesting to report in a series of embryos and juveniles.

Finally, the specimen of station 3473 from the coast of Nova Scotia has such faintly marked teeth on the suckers of the arms that it is impossible on many of the suckers to distinguish them. The specimen is 170 mm long including the tentacles, whose suckers have excessively small denticulations. A circle of 20 luminous organs is around the right eye.

Calliteuthis sp. juv.

Expedition of 1911: Stn. 3064, depth 0–2,700 m. July 21, 38°03′30″ N, 4°46′45″ E. Bourée net rapidly retrieved. Mediterranean, southeast of Minorca.

This young specimen is rather well preserved, but the species features are not clear enough to make out whether it belongs to *C. meneghinii* or *C. verrilli*. The structure of the brachial suckers indicates it is more closely related to *C. meneghinii*.

71 *Calliteuthis* ?

Expedition of 1912: Stn. 3312, depth 0–3,500 m. September 1, 45°02′ N, 13°05′ W. Bourée net rapidly retrieved.

Fragment of skin of an incomplete cephalopod, probably a *Calliteuthis* as judged from the shape of the luminous organs.

Subgenus *Stigmatoteuthis* G. Pfeffer, 1900

Calliteuthis (*Stigmatoteuthis*) *verrilli* G. Pfeffer, 1912

1881. *Calliteuthis reversa* Verrill (**83**), p. 295
1884. *Calliteuthis reversa*, Verrill (**85**), p. 243
1912. *Stigmatoteuthis verrilli* G. Pfeffer (**65**), p. 285

Expedition of 1913: Stn. 3360, depth 0–3,500 m. July 24, 47°06′30″ N, 8°11′ W. Bourée net rapidly retrieved. In the Gulf of Gascony. Stn. 3412, depth 5,270 m. August 9, 39°59′ N, 56°20′ W. Trawl. Sigsbee Trench, southwest of Nova Scotia. Stn. 3414, depth 0–4,000 m. August 10, 40°15′ N, 56°25′ W. Bourée net rapidly retrieved. Same location as above. Stn. 3447, depth 0–1,100 m. August 27, 43°13′ N, 60°11′ W. Bourée net rapidly retrieved. South of the Island of Cape Breton. Stn. 3452, depth 0–2,500 m. August 28, 43°26′ N, 59°03′ W. Bourée net rapidly retrieved. Same location as the previous station. Stn. 3453, depth 0–1,000 m. Same locality. Bourée net rapidly retrieved.

The subgenus *Stigmatoteuthis*, created by G. Pfeffer to group the species of genus *Calliteuthis* with suckers with teeth, seems to me to have only a very limited value. It leaves only the *C. meneghinii* (ex *reversa*) in the old genus *Calliteuthis* and groups all the other species under this new name. As I have already stated, I believe that the denticulation of suckers is not a valid distinction because the teeth are always too small and so variable that they may change with the age of the animal. With the present state of knowledge, one could leave Pfeffer's species in the genus *Calliteuthis* in its earlier sense. I do not have adequate evidence at hand to solve the problem further, since I do not have specimens of all the species. Chun and Pfeffer have apparently based their new names merely on the text of their predecessors and not on the study of specimens. I therefore opt for Pfeffer's provisional solution while awaiting new reports to determine the establishment of a new classification.

Pfeffer has taken his *Stigmatoteuthis verrilli* from *Calliteuthis reversa* Verrill. The latter author had given this name to several very similar cephalopods in many publications. Pfeffer re-examined these memoirs and
72 considered one of the *C. reversa* of Verrill adequately different from the earlier defined type species to merit its separation. This is how the *Calliteuthis reversa* Verrill, 1881 and 1884 became *Stigmatoteuthis verrilli* Pfeffer.

The *C.* (*St.*) *verrilli* is in fact different. Its tentacles are more slender, longer and very flattened. The suckers have a small but distinct denticulation.

The two fins meet on the median line behind the end of the body in a fleshy lamella that distinctly extends beyond this end. In *C. reversa* the end of the body has a small pointed projection between the two fins which reach the point of contact but do not extend beyond it.

This species can be rather easily recognized by these not very striking features.

This batch of cephalopods contains 19 individuals, mainly in good condition. The measurements were as follows: 95 to 142 mm in nine specimens including the tentacles, and 40 to 100 mm in ten specimens without tentacles.

The proportion of the tentacles is rather variable, the sizes of the animals differ less than the figure for their total lengths suggests.

The cephalic asymmetry discussed earlier in the text was detected in all these animals. There is a circle of 18 to 20 luminous organs around the right eye which is smaller and normal in appearance.

I observed an interesting detail on the mantle luminous organs of the one specimen from station 3452. Almost all these organs are on the ventral side; however, some are present on the dorsal side. All these organs contain an upper convex lens directed toward the tip of the arm, and a circular lens perpendicular to the first in the plane of the skin. It is encased in a black pigmented frame; the back of this lens is elliptical, much larger than its counterpart on the ventral side, and surrounded by such a thin pigmented frame that it seems like very fine piping.

A horizontal line of large luminous organs is above the funnel on the lower margin of the head. The middle organ is exactly between the two ligaments attaching the funnel to the head. To the right, toward the smaller right eye, there are five or six luminous organs; to the left there are only two or three which do not reach the eye which is large with an irregular eyelid enclosing only a small number of luminous organs, which are scattered at random.

The tentacular club has dispersed suckers, as described by Verrill. They have a large number of pointed, short, very small denticulations. Sometimes they are so small that their presence might be doubted. This feature is thus rather variable, at least in respect to size of these teeth, and it is not astonishing that they have passed unnoticed.

The globular suckers of the arms also have teeth but their size varies. These teeth are crenulated, extremely
73 low and occur on only half of the horny rim. In some specimens they were so low and separated from each other by such faint indentations that I frequently doubted their presence. This then is the distinguishing feature between *C. meneghinii* and *C. verrilli*, subject to variations that strongly detract from its specific importance.

In all these specimens the fin, as already mentioned, extends far beyond the posterior end of the body. The termination of the gladius can be clearly detected below the skin between these two fins.

I successfully removed an intact gladius from one specimen from station 3414. This was a large individual more than 140 mm long.

Genus *Histioteuthis* d'Orbigny, 1839

Histioteuthis bonelliana Férussac, 1835

1835. *Cranchia bonelliana* Férussac (**18**), *Mag. de Zool.* v.
1839. *Histioteuthis bonelliana*, Férussac and d'Orbigny (**60**), p. 327
1851. *Histioteuthis rüpellii* Vérany (**80**), p. 114
1879. *Histioteuthis collinsii* Verrill (**83**), p. 241 of the journal, 1879

1885. *Histiopsis atlantica* Hoyle (**23**), p. 201
1886. *Histiopsis atlantica*, Hoyle (**24**), Chall, p. 180
1910. *Histioteuthis rüppellii*, Chun (**14**), p. 147
1912. *Histioteuthis bonelliana*, Pfeffer (**65**), p. 297

Expedition of 1912: Stn. 3279, depth 0–3,000 m. August 23, 38°55′ N, 34°07′30″ W. Bourée net rapidly retrieved. Azores, east of Flores Island. Stn. 3281, depth 0–4,000 m. August 24, 39°25′40″ N, 35°14′30″ W. Bourée net rapidly retrieved. Same location. Stn. 3284, depth 0–1,000 m. August 24, 39° 19′ N, 35°24′30″ W. Bourée net rapidly retrieved. Same location. Stn. 3285, depth 0–500 m. August 24, 39°23′ N, 35°18′30″ W. Bourée net rapidly retrieved. Same location.

Family GONATIDAE Hoyle, 1886

Genus *Gonatus* Gray, 1849

Gonatus fabricii (Lichtenstein), 1818

1818. *Sepia loligo* Fabricius (**18**), p. 358
1818. *Onychoteuthis fabricii* (Lichtenstein) (**52**), p. 13
1843. *Onychoteuthis amoena* Möller (**56**), p. 3
1849. *Owenia megalops* Prosch (**66**), Pl. I, Fig. 7
1880. *Lestoteuthis kamtschatica* Verrill (**83**), p. 251
1881. *Gonatus fabricii*, Steenstrup (**75**), p. 9
1881. *Cheloteuthis rapax* Verrill (**82**), p. 293
74 1881. *Lestoteuthis fabricii*, Verrill (**83**), p. 291
1886. *Gonatus fabricii*, Hoyle (**25**), p. 41
1898. *Gonatus antarcticus* Lönnberg (**54**), p. 51
1912. *Gonatus fabricii*, G. Pfeffer (**65**), p. 230

Expedition of 1912: Stn. 3318, depth 0–4,500 m. September 2, 46°14′30″ N, 10°10′30″ W. Bourée net rapidly retrieved. In the middle of the Gulf of Gascony.

Expedition of 1913: Stn. 3453, depth 0–1,000 m. August 28, 43°26′ N, 59°03′ W. Bourée net rapidly retrieved. Southeast of Sable Island (Nova Scotia).

The specimen from station 3318 is in very poor condition; only the head and half the body without the skin remain. Its total length, including tentacles, could be 200 to 220 mm.

Two specimens were caught at station 3453, one 120 mm and the other 85 mm in total length. They are both completely without skin.

Geographic distribution. Various localities of the Polar Atlantic, Kamschatka, north coast of the United States, Strait of Magellan, Cape of Good Hope, the Mediterranean, Gulf of Gascony and Japan.

Family BENTHOTEUTHIDAE G. Pfeffer, 1912

Genus *Benthoteuthis* Verrill, 1885

Benthoteuthis megalops Verrill, 1885

1885. *Benthoteuthis megalops* Verrill (**86**), p. 402

1885. *Bathyteuthis abyssicola* Hoyle (**23**), vol. 1, p. 272
1900. *Bathyteuthis abyssicola*, G. Pfeffer (**62**), p. 73
1910. *Benthoteuthis megalops*, Chun (**14**), p. 185
1912. *Benthoteuthis megalops*, G. Pfeffer (**65**), p. 325

Expedition of 1911: Stn. 3131, depth 0–3,500 m. August 16, 35°09′ N, 21°21′ W. Bourée net rapidly retrieved. Azores.

Expedition 1912: Stn. 3202, depth 0–3,000 m. August 5, 31°45′ N, 20°17′ W. Bourée net rapidly retrieved. Near Madeira.

A young individual, with a total length of approximately 70 mm including the tentacles; in very poor condition.

The specimen from station 3131 is almost intact except that it has lost one tentacle. Its total length is 135 mm including the tentacle which is 88 mm itself. The body, with arms, is 55 mm. It is brick-red with a darker head, as shown by Chun. But we do not see the small white pearls shown by the author at the bases of the arms; they are replaced here by slightly darker masses of sombre-colored chromatophores.

75 *Geographic distribution.* American North Atlantic, Benguela Current, North Equatorial Current of the Indian Ocean, the Mediterranean, Marion and Crozet Islands, Cape Mala, Azores and Madeira.

Genus *Tracheloteuthis* Steenstrup, 1882

Tracheloteuthis riisei Steenstrup, 1882

1882. *Tracheloteuthis riisei* Steenstrup (**76**), p. 294
1882. *Tracheloteuthis behnii* Steenstrup (**76**), p. 294
1884. *Entomopsis velaini* Rochebrune (**67**), p. 21
1884. *Entomopsis clouei* Rochebrune (**67**), p. 21
1884. *Verrilliola gracilis* G. Pfeffer (**61**), p. 22
1884. *Verrilliola nympha* G. Pfeffer (**61**), p. 23
1910. *Brachioteuthis riisei*, Chun (**14**), p. 8
1912. *Tracheloteuthis riisei*, G. Pfeffer (**65**), p. 355

Expedition of 1912: Stn. 3202, depth 0–3,000 m. August 5, 31°45′ N, 20°17′ W. Bourée net rapidly retrieved. Near Madeira.

This small specimen, 3 cm in total length, is in very bad condition. The head is separated from the body and has no fins. It is, however, recognizable by its arms which are identical to those that G. Pfeffer has shown in Figs. 1 and 2 of his Pl. 16.

Expedition of 1913: Stn. 3439, depth 0–1,000 m. August 26, 42°44′ N, 62°56′ W. Bourée net rapidly retrieved. South of Halifax. Three whole individuals and two separate heads. The former three are complete and in good condition. They agree well with the type species and do not have noticeable peculiarities. However, a very accentuated golden pigmentation of the iris is observed. Stn. 3453, depth 0–1,000 m. August 28, 43°26′ N, 59°03′ W. Bourée net rapidly retrieved. Localities south of the Island of Cape Breton.

The second specimen is in better condition than the preceding one from station 3202, but it is without fins. It corresponds to Fig. 5 of Pl. 26 by G. Pfeffer. Its tentacles are identical with those described by this author.

Geographic distribution. North Atlantic, Portugal, Morocco, Mediterranean, extreme South Atlantic, Zanzibar, Saint Paul Island, Equatorial Pacific, Chile and Tierra del Fuego.

Family OMMATOSTREPHIDAE Steenstrup, 1861

Genus *Ommatostrephes* d'Orbigny, 1835

Ommatostrephes sagittatus (Lamarck), 1799

Expedition of 1912: Stn. 3213, depth 0–500 m. August 7, 30°45′40″ N, 25°47′ W. Bourée net rapidly retrieved. Night catch. Strait of Madeira. One specimen with a total length of approximately 20 cm, no trace
76 of a hectocotylus, in bad condition. Stn. 3279, depth 0–3,000 m. August 23, 38°55′ N, 34°07′30″ W. Bourée net rapidly retrieved. Azores. A specimen in good condition with a total length of approximately 5 cm. Stn. 3284, depth 0–1,000 m. August 24, 39°19′ N, 35°24′30″ W. Bourée net rapidly retrieved. Night catch. Azores. Specimen with total length of 27 cm, without hectocotylization; in rather poor condition.

Geographic distribution. European North Atlantic, polar Arctic Ocean, Mediterranean, Madeira and Azores. S. Berry (Rew. Cep. N. W. AM. p. 294) reports finding three specimens from the waters around Victoria (British Columbia) by Dawson. He believes this species corresponds to the subspecies *sloanei* (= *pacificus* Steenstrup) which abounds in Japan.

Genus *Illex* Steenstrup, 1880

Illex Illecebrosus (Lesueur), 1821

1821. *Loligo illecebrosa* Lesueur (**51**), p. 25
1837. *Loligo coindetii* Vérany (**78**), p. 94
1839. *Ommastrephes sagittatus* Férussac and d'Orbigny (**60**), p. 345
1851. *Loligo sagittata* Vérany (**80**), p. 110
1880. *Illex illecebrosus* Steenstrup (**74**), p. 82
1880. *Illex coindetii*, Steenstrup (**74**), p. 83
1900. *Illex illecebrosus*, G. Pfeffer (**62**), p. 179
1912. *Illex illecebrosus illecebrosus*, G. Pfeffer (**65**), p. 405
1912. *Illex illecebrosus coindetii*, G. Pfeffer (**65**), p. 393

This list of synonymy is only a brief look at the principal names under which this cephalopod has been designated. Many other authors have also used names oscillating among the three species, *coindeti*, *illecebrosus* and *sagittatus*.

Expedition of 1913: Stn. 3452, depth 0–2,500 m, August 28, 43°26′ N, 59°03′ W. Bourée net rapidly retrieved. East coast of the United States.

Four specimens of this species were captured in good condition. Two have a total length of 39 cm including the tentacles, the third is 41 cm and the fourth is 42 cm long.

G. Pfeffer has divided the numerous specimens reported by the authors into two varieties named *Illex illecebrosus illecebrosus* for the American and *Illex illecebrosus coindeti* for the European forms. He was able to verify that the species briefly described and illustrated by earlier authors agree well with his species. However, I do not think this subdivision into two geographic varieties is justified. It would be more appropriate to designate the names of these cephalopods as follows: *Illex illecebrosus* (Lesueur) var. *illecebrosus* G. Pfeffer and *Illex illecebrosus* (Lesueur) var. *coindeti* G. Pfeffer.

The specimens I examined correspond well with the description given by G. Pfeffer for the American variety. None of them show any signs of dorsal arm modification for hectocotylization.

77 *Geographic distribution.* European coasts of the temperate Atlantic, Mediterranean, Atlantic coast of the United States and Brazil.

Genus *Stenoteuthis* Verrill, 1880

Stenoteuthis bartrami Lesueur, 1821

1821. *Loligo bartrami* Lesueur (**51**), p. 90
1835. *Omastrephes bartrami*, Férussac and d'Orbigny (**60**), p. 55
1880. *Stenoteuthis pteropus* Verrill (**83**), p. 228
1900. *Stenoteuthis bartrami*, G. Pfeffer (**62**), p. 180
1912. *Stenoteuthis bartrami*, G. Pfeffer (**65**), p. 465

Expedition of 1911: Stn. 3126, surface. August 14, 32°34′45″ N, 17°05′30″ W. Deserta Grande Island near Madeira.

The specimens were caught in a shrimp net under artificial light.

One specimen is 29 cm long, another 32 cm long and a third 28 cm, caught by spearing.

These three cephalopods are in good condition and do not show any arm modification for a hectocotylus.

Geographic distribution. Entire Atlantic. Some parts of the Pacific, Mediterranean, Marshall Islands, Fiji and Bering Sea.

Stenoteuthis pteropus (Steenstrup), 1855

1855. *Ommatostrephes pteropus* Steenstrup (**71**), p. 200
1880. *Stenoteuthis megaptera* Verrill (**84**), p. 223
1900. *Stenoteuthis pteropus*, Pfeffer (**62**), p. 181
1912. *Stenoteuthis pteropus*, Pfeffer (**65**), p. 490

Expedition of 1911: Stn. 3126, surface. August 14, 32°35′45″ N, 17°05′30 W. Deserta Grande Island, near Madeira.

One large specimen captured by spearing under electric light.

Geographic distribution. European and American North Atlantic, Mediterranean, Indian Ocean and Antilles.

Genus *Rhynchoteuthion* G. Pfeffer, 1908

1903. *Rynchoteuthis* Chun (**9**), p. 716
1908. *Rynchoteuthis* Pfeffer (**63**), p. 88

Expedition of 1910: Stn. 3053, surface. September 13, 37°20′ N, 0°07′ W. Fine-meshed net. Southwest of the Balearic Islands.

Expedition of 1911: Stn. 3097, surface. August 1, 30°40′ N, 34°15′ W. Fine-meshed net. Near the Grand Salvages.

78 Expedition of 1913: Stn. 3383, surface. August 4, 38°40′ N, 34°15′ W. Fine-meshed net, speed 10 knots, from 7 to 7:30 p.m. Azores.

With the present level of knowledge, this genus *Rhynchoteuthion* is known only as the larvae form of Ommatostrephidae. However, it has not yet been ascertained whether all the members of this family pass through this stage. Nor has it been found to which adult species these observed individuals would belong.

The first name given by Chun in 1903 was *Rhynchoteuthis*; in 1908 G. Pfeffer changed it to *Rynchoteuthion* since the former genus was already in use for a fossil cephalopod.

The 1911 specimen is in such poor condition and so contracted that the entire head had retracted in the mantle, showing only two united arms. It corresponds to the stage represented by Chun in Fig. 8 of his Pl. 28.

The 1910 and 1913 specimens are similar in size, but their ventral arms are longer than those of the 1911 individual. They correspond to Figs. 6 and 7 of Pl. 28 by Chun.

Family CHIROTEUTHIDAE Gray, 1849

The composition of this family has undergone various transformations following the continuous discovery of new species and larval forms. This has resulted in confusion, uncertainty and considerable nomenclatural complications.

In 1912, G. Pfeffer tried to bring order into this chaos and achieved an initial success. I shall follow the first part of his classification but will radically modify the second part.

He divides the Chiroteuthidae into two large sections based on the presence of four rows of suckers, or more than four rows on the tentacular club. This is the most striking character; I put to the side the other features, particularly those that relate to the luminous organs.

The first section with four rows of suckers is Chiroteuthinae and the other is Mastigoteuthinae. Let us leave the latter for the moment.

G. Pfeffer places only the genus *Chiroteuthis* d'Orbigny, 1839 in Chiroteuthinae, but he divides it into two sections. The first includes the larval or immature forms which constitute the earlier genera *Doratopsis* and *Planctoteuthis*. The second includes the adult forms where he places the subgenera *Chiridioteuthis*, *Chirothauma* and *Chiroteuthis sensu stricto*.

The first division is open to criticism. Either *Doratopsis* and *Planctoteuthis* are larvae and thus do not merit a special name or they are adults and have no place in the genus *Chiroteuthis* from which they differ greatly, and their particular name must be preserved.

79 Pfeffer's suggestion is also the most reasonable, that *Doratopsis* and *Planctoteuthis* are the larvae of *Chiroteuthis*. This seems correct and I support the view. The names of these two genera must be completely suppressed, keeping only the earlier name and synonymizing the other. One must not write the name *Doratopsis vermicularis* Rüppell like Pfeffer in the following manner: *Chiroteuthis* (*Doratopsis*) *vermicularis* Rüppell. It is one or the other name but not both. According to me, it would be correct to write *Chiroteuthis vermicularis* (Rüppell) and in the synonymy write *Doratopsis vermicularis* Rüppell indicating that it is the young form.

The following point also deserves criticism: Pfeffer considered genus *Chiroteuthis* to be two forms; one *sensu lato* and the other *sensu stricto*, so that we have a large genus *Chiroteuthis* containing three subgenera, one of which is the small subgenus *Chiroteuthis*. Because of this, he is forced to write *Chiroteuthis* (*Chiroteuthis* s. str.) *veranyi* Férussac for *Chiroteuthis veranyi* Férussac 1835. This is a serious impediment to binomial nomenclature.

However, he goes still further. He believes that this *C. veranyi* has two varieties: one European, the other American. The European form is the strictly *C. veranyi* and the American form is the *C. lacertosa* of Verrill and he qualifies the first as *Chiroteuthis* (*Chiroteuthis* s. str.) *veranyi veranyi* Férussac, 1835 and the second as *Chiroteuthis* (*Chiroteuthis* s. str.) *veranyi lacertosa* Verrill, 1881. We are now even farther from binomial writing. If we wish to push the matter to its limit, which Pfeffer did not do, we can arrive at the following situation: a particular genus *Leptoteuthis* of Verrill is probably the young form of his *Chiroteuthis lacertosa*, but to continue the nomenclature as just used for *Doratopsis* we would have to write: *Chiroteuthis* (*Chiroteuthis* s. str.) *Leptoteuthis veranyi lacertosa* Verrill. It cannot be said that this method simplifies and clarifies the problem.

To sum up, I suggest the suppression of the genera *Doratopsis*, *Planctoteuthis* and *Leptoteuthis* since they are the young forms of *Chiroteuthis*; they must be completely synonymized. The two subgenera *Chirothauma* Chun and *Chiridioteuthis* Pfeffer have only the value of species in the genus *Chiroteuthis* d'Orbigny. Finally, *Chiroteuthis veranyi* Férussac differs enough from *Chiroteuthis lacertosa* Verrill for the two species to be kept separate.

From the foregoing it can be seen that the composition of the subfamily of Chiroteuthinae Pfeffer must include the following species:

Genus *Chiroteuthis* d'Orbigny

1835. *Chiroteuthis veranyi* (Férussac), adult forms
1912. *Chiroteuthis portieri* Joubin, adult forms
1881. *Chiroteuthis lacertosa* Verrill, adult forms
1894. *Chiroteuthis picteti* Joubin
1896. *Chiroteuthis macrosoma* Goodrich
80 1896. *Chiroteuthis pellucida* Goodrich
Chiroteuthis vermicularis Rüppell, larval forms
1908. *Chiroteuthis sagitta* Chun, larval forms
Chiroteuthis richardi Joubin, larval forms
1908. *Chiroteuthis exophthalmica* Chun, larval forms
1908. *Chiroteuthis lippula* Chun, larval forms
1912. *Chiroteuthis planctonica* Pfeffer, 1912
1884. *Chiroteuthis diaphana* Verrill, 1884

It goes without saying that the larval forms are destined to disappear from this list for they will be affiliated either to already known adults or they constitute the young of adults yet to be discovered. In the former case they will come into synonymy and in the latter case their species name will remain in the genus *Chiroteuthis* and it will be the adult that will assume the species name of the larva.

Subfamily Chiroteuthinae G. Pfeffer, 1912

Genus *Chiroteuthis* d'Orbigny

Chiroteuthis portieri, Joubin 1912
(Pl. XIII, Figs. 4–11)

1912. *Chiroteuthis portieri* Joubin (**40**), C. R. p. 395
1916. *Chiroteuthis portieri*, Joubin (**42**), no. 317

Expedition of 1911: Stn. 3100, depth 0–3,500 m. August 2, 29°03′ N, 16°08′30″ W. Bourée net rapidly retrieved. Straits of the Ilhéus Selvagens.

This cephalopod is unfortunately in very poor condition: as such, the description is very incomplete. However, due to certain special characteristics observed and the significant differences that would distance it from the other Chiroteuthidae species described so far, I believe I must erect a new species for this form that I dedicate to Professor Portier of the Institute of Oceanography.

I have already given a preliminary description of this cephalopod in a note to the Academy of Sciences in 1912 (**40**) and a more complete description in the Bulletin of the Institute of Oceanography (**42**) No. 317 in 1916.

The specimen, as a whole, is characterized by its extremely graceful eight arms, two of which are very long. This feature contrasts with the arrangement normally seen in *Chiroteuthis* and imparts a very special physical appearance to the animal. The body is elongated, narrow in its cephalic part and slightly swollen in the abdominal region. The fins are united with the terminal skin of the abdomen and follow the normal arrangement in *Chiroteuthis*.

The two eyes and the buccal bulb have come out of their respective cavities. The suckers are exceptionally small and resemble those in Mastigoteuthinae; there are a great many distributed on the arms. Although the
81 tentacles are completely missing, the animal has the general appearance of *Chiroteuthis*. Study of the details confirms this first impression. The feature chiefly distinguishing it from the species already known is the absence of the broad arms which here are as slender as the others.

The head is narrow and very elongated. A slightly marked constriction separates it from the brachial crown which is not adequately developed in proportion to the gracefulness of the arms. The plate very accurately shows this arrangement; it also indicates a deformation due to the bad condition of the specimen. A torsion that veers all the arms toward the left is observed here; it has pushed the two short arms out of a row and has produced a groove passing between the two eyes and reaching the brachial crown to the left whereas it should be in the sagittal plane. It is easy to rearrange things in place by pushing the two short arms to the right and left of the sagittal plane.

The entire head is semitransparent and the internal organs, particularly the nerve ganglia, can only be surmised. The cephalic region is 4 cm long from the level of the buccal bulb to the ventral margin of the mantle opening.

The funnel is firmly attached to the head and only its terminal tube is strongly curved and free; its lateral pillars are long; the canalicular parts of the ink canal can be seen under the lower opening. Their oval-shaped funnel locking cartilages are in the lower angles on each side; this will be discussed later in the text. I could not detect the valve in the funnel. This entire organ is weak, soft, has thin walls and gives the impression of belonging to a weak swimmer. This assessment also corresponds well with the development of the caudal fins.

It is impossible to say anything about the eyes that are empty, protuberant and flaccid. I have seen no trace of luminous organs on the covering of the globe or neighboring skin.

The funnel locking cartilage consists of a pair of grooves on each side of the funnel's fold above the pillar; it is oval-shaped. The oval opening, apically a little narrower, has a thin, wavy, transparent, membranous ridge forming a fringe to the lateral and upper margins of the groove. The opening is bordered within by opaque, pearly white ring of fibrous cartilage. The cavity is scooped from a thick transparent cartilage which protrudes below the groove in the pillar of the funnel. The entire organ is 4 mm long, almost half is occupied by the opening and the other half by the oval-shaped cartilaginous groove where the ridge is located.

The ridge on the inner edge of the mantle is shaped like a basally curved hook with a soft tip resting on a small, noticeably rectangular cartilaginous support. It is approximately 4 mm long. Thus the entire adhesive organ is small in relation to the overall size of the animal and the firmness of the pallial closure which it produces must be mediocre. This agrees well with the weakness of the funnel and the fins.

82 The mantle is unfortunately in very bad condition; it is yellowish-white, elongated, oval-shaped and slightly swollen in the anterior third. Its opening is slightly narrower than its average width. The ventral mantle margin is circular without sinuosities; on the dorsal side it forms a strong tip rising to the nuchal part where it is articulated by a cartilage with rectilinear furrows and ridges. The two small white hooks already described are on the inner face of the ventral margin.

The fins are relatively small and round; the two halves are separated by the pointed termination of the abdomen, which extends into a very small tubercle between the fins. It is torn and, therefore, it is impossible to assess its exact proportions. Moreover, this entire portion of the body is macerated and so soft that it is difficult to find the interrelations of the parts. This small median tubercle was probably the beginning of a more developed appendage. On the dorsal face, the fins rise slightly on each side of the dorsomedian line and soon disappear. The edge of the fins is thin, transparent and slightly wavy.

The sheath of the gladius is visible through the transparent mantle but it is torn and at the top only the small, brown, filiform tip remains. Judging from the shape of the mantle it seems to be enlarged at the level of the upper insertion of the fins.

The arms are characterized by the great length of the three dorsal and lateral pairs, by the slenderness of the four pairs and the poor development of their muscles. They gradually and very slightly become attenuated from the head onward, becoming filiform in their terminal part. Their base is attached to the head by a rudimentary brachial membrane which forms a very slightly projecting crown; they are thus almost completely independent of each other. Their cross section is approximately triangular with the base corresponding to the oral surface of the arm; the apex of the triangle forms an external keel which is well-marked only in the basal part of the arms and soon disappears. The lower third of the inner face of the arms and especially around the mouth is pigmented purplish-red. This hue becomes fainter with increasing distance from the mouth.

All the arms are worn out, their extreme tips are missing over a more or less greater length, and some have only the muscles. I found no trace of luminous organs on these arms.

The lengths of the fourth pair of the ventral arms, are only 76 and 68 mm; they are very much shorter than the others; their cross section is also more clearly rectangular and the base is more brightly colored, even on the outer side, than in the other arms. The first centimeter has no suckers; then at first they are widely spaced, alternating to the right and left on the two edges defining the oral surface of the arms. Thereafter, they are more closely arranged. Commencing from about one-third of the length, two new rows of suckers
83 become intercalated between the previous two, consequently on the oral face of the arms forming four rows of suckers. Near the middle two additional rows appear, bringing the total to six. The suckers are set very close to each other and it is very difficult to decipher their order. They form six longitudinal parallel rows, but not being at the same level they appear to be arranged in oblique transverse rows with six suckers in each. The oral surface of the arm has to be examined with great attention to detect this double symmetry because the suckers are very small and, since they also have long peduncles, they become entangled. This arrangement of six rows is seen only in the middle of the arm; soon there are only four rows, and then no more than two in the last fourth of the arm. At the end, since the arm is very slender, the ridges are very close together and the suckers seem to merge into a single row.

The middle arms are very much longer than the ventral arms. The third pair exceeds 30 cm, its base is slightly swollen but is rapidly diminishes to its average size and slender appearance; it becomes filiform only at the very end. The suckers start at the base very close to the buccal bulb and are arranged in four rows. Two centimeters farther, six rows continue to the last fourth where there are no more than four rows. Only at the tip are there only two rows. This pair of arms does not seem to have had a keel, but it is in such bad condition that I could not confirm this.

The middle second pair of arms also reaches 30 cm; the suckers in these are arranged as in the third pair. The keel of the outer or aboral surface can be better detected at the base.

The first dorsal pair is a little less than 30 cm long and identical in size, shape and arrangement of the suckers with the two preceding pairs. It is, therefore, useless to elaborate this point.

All these arms, highly independent of each other due to the insignificant size of the interbrachial membrane, show, however, for 2 to 3 cm at their base, a very slight lateral membranous crest, an extension of this small web.

I can say nothing about the tentacles. Only one filiform fragment, a few millimeters in length and reduced to its muscular axis, remains attached in the normal place.

The suckers are extremely small; each rests on a small pad of skin at the top of which a very slender muscular peduncle resembling a silken thread is implanted. The yellow and brown horny sucker is at its tip. There are several hundred on each arm. They are all approximately of the same size except at the filiform
84 extremity of the arms where they shorten to almost half their length. They are in all from 2 to 3 mm long including the peduncle. The horny part measures, on an average, 0.33 mm.

The horny, thimble-shaped cup [ring] of the sucker has a large circular opening on its side. There are teeth on the lateral margins of the opening. The two middle teeth furthest from the peduncle are pointed and slightly hooked; the others are soft. The largest tooth on each side is closest to the pointed tooth. The others gradually diminish toward the base. There are four soft teeth on each side. The posterior edge of the circular opening has minute undulations too reduced to qualify as teeth. The posterior third of the sucker opening is screened by a very thin, transverse horny membrane like a transparent veil, slightly sunken within the sucker. The teeth and the ring on which they rest are dark brown. The cup of the sucker is convex, smooth and has no ornamentation or protrusions like those found on the suckers of other representatives of the same family.

The opening at the base of the sucker is broad and the flare of the peduncle fits into it. The circumference of the orifice is posteriorly rounded but has a depression below the lateral opening. The peduncle is slender and transparent, allowing one to see a double system of muscle fibers twining in two opposite spirals in the interior of the sucker where they form the piston. I examined a large number of suckers; they were all identical and differed only in insignificant details of teeth.

The buccal bulb was severed from its cavity; I could assess that it is very small, but I could not see the arrangement of the lips.

I am left with giving the measurements taken from this cephalopod (in mm).

Length of the head		..	40
Length of the abdomen (ventromedial line)		..	55
Length of the abdomen (dorsomedial line)		..	67
Length of the fins		..	15
Width of the fins		..	17
1st dorsal arm, left (incomplete) 286	right (incomplete)	..	284
2nd lateral arm, left (incomplete) 292	right (incomplete)	..	223
3rd lateral arm, left (incomplete) 308	right (incomplete)	..	272
4th ventral arm, left (incomplete) 76	right (incomplete)	..	68

Despite the absence of the tentacles and the generally poor condition of the specimen, it can be concluded that the animal belongs to the genus *Chiroteuthis*. This cephalopod seems to be a pelagic form with slow movements because of the weakness of its natatory musculature. It must feed on microplankton because
85 of the weakness of its suckers and buccal bulb. It must capture the plankton by slowly moving its arms around itself. The absence of luminous organs makes me think that it lives at shallow depths.

Chiroteuthis veranyi Férussac, 1835

1835. *Loligopsis veranyi* Férussac (**17**), pl. 65
1839. *Chiroteuthis veranyi*, Férussac and d'Orbigny (**60**), p. 325
1851. *Loligopsis veranyi*, Vérany (**80**), p. 120
1885. *Chiroteuthis veranyi*, Hoyle (**23**), p. 330
1894. *Chiroteuthis veranyi*, Joubin (**31**), p. 1
1910. *Chiroteuthis veranyi*, Chun (**13**), p. 240
1912. *Chiroteuthis veranyi veranyi*, G. Pfeffer (**65**), p. 594

Expedition of 1912: Stn. 3217, depth 0–1,000 m. August 8, 30°47′ N, 25°47′ W. Bourée net rapidly retrieved. Azores.

The sole specimen captured is a complete adult in good condition. Some interesting measurements (mm) are:

Length of the fin	..	20
Length of the body	..	42
Length from the tip of the fin to the base of the arms	..	70
Length from the posterior end of the body to the tip of the ventral arms	..	185
Length of tentacles	..	350

This cephalopod conforms well to the type specimen described by Férussac. It was further qualified by Pfeffer as *Ch. veranyi veranyi* and there is no possibility to confuse it with the *Ch. lacertosa* of Verrill.

Geographic distribution. Mediterranean and Northeast Atlantic.

Chiroteuthis picteti Joubin, 1894

1894. *Chiroteuthis picteti* Joubin (**31**), p. 40
1896. *Chiroteuthis macrosoma* Goodrich (**21**), p. 12
1900. *Chiroteuthis picteti*, G. Pfeffer (**62**), p. 186
1903. *Chiroteuthis picteti*, Chun (**10**), p. 67
1906. *Chiroteuthis picteti*, Nishikawa (**58**), p. 109
1909. *Chiroteuthis* (*Chirothauma*) *imperator* Chun (**14**), p. 241
1912. *Chiroteuthis* (*Chirothauma*) *imperator*, G. Pfeffer (**65**), p. 81
1916. *Chiroteuthis imperator* Sasaki (**75**b), p. 108

Expedition of 1911: Stn. 3084, depth 0–3,000 m. July 28–29, 34°20′30″ N to 33°46′ N, 10°05′ W to 10°44′ W. Small Bourée net. Between Gibraltar and the Ilhéus Selvagens. Stn. 3086, depth 0–4,300 m. July 29, 33°46′ N, 10°44′ W. Bourée net rapidly retrieved. Same locality.

Geographic distribution. Amboina, Japan, Sumatra, Indian Ocean, Atlantic Ilhéus Selvagens.

86 This splendid cephalopod is known only from a very small number of specimens. The two which served for me as the type species in 1894 were collected by Messrs. Bedot and Pictet from Ambiona. Three specimens were examined by Chun (one from Sumatra and two from Japan). The specimen reported by Goodrich was collected from the Indian Ocean and those by Nishikawa and Sasaki were from Japan.

I had dedicated this new species to Pictet of Geneva in 1894. Chun first accepted this name in 1903, but in 1909 and in 1910 he considered it necessary to change it to *Chiroteuthis imperator*. There is no basis for this change since the small bump he reported on the peduncles of the suckers and which seemed to be his best argument, is so insignificant that it is not even enough to justify the establishment of a variety. Obviously, dedicating such a beautiful cephalopod, depicted in magnificent plates, to the emperor is a tempting flattery. Nevertheless, my species name has priority and my watercolor sketch, though less artistic, is characteristic and demonstrative. Among the authors subsequent to my publication, some, like Goodrich in 1896, created a new species (*macrosoma*), not knowing mine, and others, like Chun in 1900 and G. Pfeffer, replaced *picteti* by *imperator*. Sasaki did the same in 1916. I believe I must re-establish the truth and the priority.

Chiroteuthis lacertosa Verrill, 1881
(Pl. XIII, Figs. 2–3)

1881. *Chiroteuthis lacertosa* Verrill (**83**), p. 408
1912. *Chiroteuthis* (*Chiroteuthis*) *lacertosa*, G. Pfeffer (**65**), p. 607

Expedition of 1913: Stn. 3439, depth 0–1,000 m. August 26, 42°44′ N, 62°56′ W. Bourée net rapidly retrieved. South of Halifax.

At this station a magnificent, complete specimen was caught in very good condition. This specimen undoubtedly agrees completely with the description of *Chiroteuthis lacertosa* Verrill.

As has been mentioned above, G. Pfeffer regarded this cephalopod as an American variety of *Chiroteuthis veranyi* recorded on the coasts of Europe. On the contrary, I think that the two forms differ enough in total aspect, and in a large number of details, as Verrill had already indicated, to justify their separation.

The drawing given by Verrill is slightly schematic and does not give an adequate picture of the animal. I thought that it would be a good idea to take advantage of the capture of a very beautiful, whole and intact specimen to make a good water-color sketch. Miss Vesque has executed one which is reproduced in Pl. XIII, Fig. 2.

The measurements (mm) of the specimen are:

87

Total length including the tentacles	. .	700
Total length without the tentacles	. .	410
Length of the tentacles	. .	550
Length of the tentacular club	. .	48
Length of the 1st arm (dorsal)	. .	120
Length of the 2nd arm	. .	142
Length of the 3rd arm	. .	206
Length of the 4th arm (ventral)	. .	270
Dorsal length of the mantle	. .	130
Ventral length of the mantle	. .	120
Length of the fin	. .	57
Width* of the fin	. .	50

I have also given an illustration of the tentacular club (Pl. XIII, Fig. 3) which shows the details of this very remarkable part better than Verrill's drawing.

Subfamily Mastigoteuthinae G. Pfeffer, 1912

Genus *Mastigoteuthis* Verrill, 1881

The species comprising this genera are extremely difficult to distinguish from one another. This is due to the inadequacy of the descriptions and illustrations given by the authors. Moreover, it must also be noted that the specimens described were almost always in bad condition. Nearly all the specimens I have examined are practically devoid of skin, eliminating the luminous organs indispensable for an accurate identification. Their absence raises doubts regarding the validity of the names assigned.

The publications of G. Pfeffer summarized completely our knowledge of the genus in 1912. This incorporated an attempt at classification, based mainly on the relative length of the fins. This criterion is definitely inadequate because it becomes difficult to determine clearly whether the fin is 4/5 or 3/4 the length of the mantle, considering that the measurements are frequently taken on single specimens contracted by the preservative fluid and more or less without skin.

It must also be taken into account that the shape and relative size of the fins are subject to variation with the age and perhaps the sex of the animal. This is frequently the case in cephalopods and it probably is the same in *Mastigoteuthis*.

This brief preamble should show that the current classification of genus *Mastigoteuthis* must be considered provisional, destined to be improved upon when a more significant series of better-preserved specimens of various ages are obtained.

88 *Mastigoteuthis grimaldii* (Joubin), 1895

1895. *Chiroteuthis grimaldii* Joubin (**32**), p. 38

*In the French original given as "length"—General Editor.

1900. *Chiroteuthopsis grimaldii*, (G. Pfeffer) Joubin (**62**), p. 187
1904. *Mastigoteuthis dentata* Hoyle (**27**), p. 34
1907. *Chiroteuthopsis grimaldii*, Joubin (**39**), p. 345
1908. *Mastigoteuthis flammea* Chun (**12**), p. 88
1910. *Mastigoteuthis flammea*, Chun (**14**), p. 229
1912. *Mastigoteuthis flammea*, G. Pfeffer (**65**), p. 619

Expedition of 1910: Stn. 3015, depth 0–4,200 m. August 30, 41°02′ N, 10°20′ W. Bourée net rapidly retrieved. In the open waters of Portugal. Stn. 3028, depth 0–5,100 m. September 4, 36°54′30″ N, 11°49′ W. Bourée net rapidly retrieved. Strait of Gorringe Bank. Stn. 3036, depth 0–4,740 m. September 7, 36°06′40″ N, 10°18′ W. Bourée net rapidly retrieved. Entrance to the Strait of Gibraltar.

Expedition of 1911: Stn. 3078, depth 0–2,000 m. July 27, 35°18′15″ N, 7°52′ W. Bourée net rapidly retrieved. Strait of Gibraltar. Stn. 3086, depth 0–4,300 m. July 29, 33°46′ N, 10°44′ W. Bourée net rapidly retrieved. Between Gibraltar and Selvagen Grande. Stn. 3131, depth 0–3,500 m. August 16, 35°09′ N, 21°21′ W. Bourée net rapidly retrieved. Azores.

Expedition of 1912: Stn. 3281, depth 0–4,000 m. August 24, 39°25′40″ N, 35°14′30″ W. Bourée net rapidly retrieved. Azores.

Expedition of 1913: Stn. 3415, depth 0–2,000 m. August 10, 40°15′ N, 56°17′ W. Bourée net rapidly retrieved. Sigsbee Trench (Cape Breton). Stn. 3452, depth 0–2,500 m. August 28, 43°26′ N, 59°03′ W. Bourée net rapidly retrieved. South of Cape Breton Island. Stn. 3526, depth 0–2,000 m. September 29, 38°30′ N, 34°40′ W. Large Richard net. Azores, east of Flores Island.

I think it is necessary to unite the following three species of genus *Mastigoteuthis* into one species: *M. grimaldii* Joubin, 1895, *M. dentata* Hoyle, 1904 and *M. flammea* Chun, 1908.

Hoyle's description of his new species *M. dentata* is inadequate. He only cites the following as a precise character, permitting the appointment of other specimens to his species. He gives the measurements of two specimens, one male and one female, that he examined, and states that: 1. *M. dentata* differs from *M. agassizi* Verrill because the latter species has horny rings on the brachial suckers without teeth, whereas *M. dentata* has teeth; 2. it differs from *M. levimana* Lönnberg by the fact that the ventral arms of *M. dentata* are thicker and not as slender as the others; however, the teeth of the suckers are similar in the two species; 3. he notes that there is an interesting difference in the shape and size of the fins between the male and female.

Two illustrations of the brachial sucker and two others giving the topography of the visceral organs form must be added to these descriptions.

89 He makes no reference to the structure of the skin, its appearance or to the presence or absence of luminous organs. However, Hoyle cites only the structure of the horny ring of the suckers as the distinguishing feature between *M. dentata* and *M. agassizi* Verrill. Now *M. agassizi* is covered with luminous organs and it must therefore be concluded that *M. dentata* is also provided with them. But Chun has compiled a dichotomous key in which he attempts to define the differences between the species of genus *Mastigoteuthis* and makes a special category for *M. dentata* based on the absence of luminous organs. On the contrary, Hoyle's text suggests that these are present.

From this incomplete description by Hoyle and his illustrations, particularly of the denticulation of the suckers and proportions of the arms, I conclude that his *M. dentata* does not differ from *M. grimaldii* Joubin.

With regard to *M. flammea* Chun, I unite it with the two preceding forms; its shape, color, size and fins are identical with theirs. The only feature separating them is the following: *M. dentata* and *M. grimaldii* have approximately one dozen denticulations on the horny ring of the brachial suckers. According to Chun,

M. flammea has only three to five. A curious fact is that Chun has given no description or illustration of these suckers in either his text or plates. This is an unfortunate omission in the establishment of a new species. This is a mistake on the part of the author, and his report of three to five teeth on the suckers must be considered without value. It seems equally to me that too much importance should not be attached to the slight variations in the proportions of the fin; these can be attributed to age, size, sex and the preservative used. For these reasons, and following the rules of nomenclature, I unite *M. dentata* Hoyle and *M. flammea* Chun to *M. grimaldii* which has priority.

Almost all the specimens captured at various stations are in very poor condition; their tentacles and arms are broken, the skin is more or less removed and most frequently only shreds remain. I attribute this condition to the excessive speed at which the Bourée net was drawn, subjecting these animals to violent rubbing and the removal of the delicate skin. This has greatly complicated identification.

Some of the features of these specimens are given here, including the measurements of the fins since the diagnosis is based on the ratio of their length to the length of the body. The variability of this feature and hence the lack of value of this criterion can be seen.

Stn. 3015. A specimen 192 mm long (body and ventral arms) without tentacles.

Dorsal length of the body	..	68 mm
Dorsal length of the fin	..	48 mm
Width of the fins	..	54 mm

90 There are luminous organs distributed on the ventral side of the body, the ventral arms, head, and dorsal [sic] surface of the fins. It is difficult to accurately give their position and number since only carmine red shreds of skin remain.

Stn. 3828. A specimen without tentacles and almost no skin, 100 mm in length (body and ventral arms).

Dorsal length of the body	..	36 mm
Dorsal length of the fin	..	20 mm
Width of the fins	..	23 mm

There are traces of luminous organs on the ventral side, between the eyes, on the underside of the ventral arms, and on the dorsal surface of the fins.

Stn. 3036. A specimen with a total length of 133 mm including the tentacle. The skin is in large part rubbed off, one tentacle is attached and another is detached.

Length of the body and ventral arms, without tentacles	..	92 mm
Dorsal length of the body	..	36 mm
Dorsal length of the fin	..	20 mm
Width of the fins	..	22 mm

Luminous organs are present above the fins and on the ventral side of the body, head and ventral arms. The tentacles have innumerable small suckers which will be discussed later in the text.

Stn. 3078. The material contains a specimen of this cephalopod and two separated tentacles which probably belonged to it.

Length of the animal (body and arms)	..	128 mm
Dorsal length of the body	..	43 mm
Dorsal length of the fin	..	25 mm
Width of fin	..	32 mm

Sparse luminous organs are on the dorsal surface of the fins, around the funnel between the eyes, and on the underside of the ventral arms.

The separated tentacles are 117 mm long; they are cylindrical and resemble a fragment of twine, gradually narrowing to a sharp point. They have innumerable minute suckers arranged over a length of 82 mm. They begin in a narrow band on the buccal side commencing at the base of the arm. The surface covered by these increases with distance from the mouth and extends to the lateral parts of the tentacle. They converge toward the suckers of the other side until they are separated by only a narrow band without suckers. Finally, the last 15 mm of the tentacle are entirely covered with a carpet of these small brown suckers.

91 Stn. 3086. A specimen without tentacles, 212 mm long (body and ventral arms). Almost completely without skin.

Dorsal length of the body .. 78 mm
Dorsal length of the fin .. 50 mm
Width of the fins .. 46 mm

Luminous organs are sparse on the dorsal surface of the fins, ventral side of the body, base of the ventral arms and between the two eyes around the funnel.

Stn. 3131. Highly deteriorated specimen, skin almost absent, no tentacles, 70 mm in length (body and ventral arms).

Dorsal length of the body .. 30 mm
Length of the fin .. 15 mm
Width of the fins .. 16 mm

Luminous organs present on the ventral side of the body, between the eyes and on the base of the ventral arms.

Stn. 3281. First specimen in poor condition, 125 mm long (body and ventral arms) accompanied by two separated tentacles each one 175 mm long, approximately 100 mm of which are covered with microscopic suckers arranged as described above. Traces of luminous organs are present on the ventral side of the body and head.

Dorsal length of the body .. 40 mm
Dorsal length of the fin .. 24 mm
Width of the fins .. 40 mm

The poor condition of the specimen renders these measurements doubtful.

Second specimen has no arms or tentacles. Only a few measurements could be taken on this mutilated body without skin:

Dorsal length of the body .. 25 mm
Dorsal length of the fin .. 14 mm
Width of the fins .. 12 mm

This is a highly dubious identification. However, the denticulation of the suckers is identical to that of the other specimens.

Stn. 3415. A small specimen in poor condition, largely without skin from the tip of the arms. Apparently its total length is 75 mm.

Dorsal length of the body .. 28 mm
Dorsal length of the fin .. 15 mm
Width of the fin .. 15 mm

92 Traces of luminous organs are present on the ventral skin, between the eyes. Suckers denticulated. Identification doubtful.

Stn. 3452. Three rather large specimens completely without skin. One is no more than a fragment and could not be measured. Along with these specimens, there are five tentacles about 230 to 275 mm long with innumerable small suckers arranged as described above.

	1st specimen	2nd specimen
Dorsal length of the body	73 mm	60 mm
Dorsal length of the fin	45 mm	40 mm
Width of the fin	62 mm	50 mm

Some luminous organs seen on the skin at the base of the ventral arms.

Stn. 3526. A small, whole specimen in very good condition including tentacles. Suckers dentate as in the other specimens. Total length 134 mm including the tentacles, 74 mm without the tentacles.

Dorsal length of the body ..	30 mm
Dorsal length of the fin ..	17 mm
Width of the fin ..	18 mm

Luminous organs present on the dorsal surface of the fin, on the ventral side of the body above the tip of the fin, between the eyes and on the ventral base of the ventral arms. The suckers of the tentacles are identical with those already described.

It must be noted that in all these cephalopods, the horny ring of the brachial suckers has a half crown of about one dozen pointed teeth which are identical in all specimens.

The examination of these specimens, their similarities, their comparison on the whole and of their details convinces me that the three species, more or less completely described by the authors, are identical. Furthermore, a comparison of the measurements of the fins and of the body suggests that the differences are due to age. In this species, as in other cephalopods, the fin changes its shape and proportions with age; sexual differences may also bring about this change. These various considerations indicate that the feature given by Chun and above all by Pfeffer regarding the proportion of the fins to differentiate the species is untenable and there is not much worth to attribute to their tables, they are only supplementary information.

Geographic distribution. Azores, Morocco, Guinea Current, St. Thomas Island, Galapagos, Cape Mala.

93 *Mastigoteuthis cordiformis* Chun, 1908

1908. *Mastigoteuthis cordiformis* Chun (**12**), p. 88
1910. *Mastigoteuthis cordiformis*, Chun (**14**), p. 222
1912. *Mastigoteuthis cordiformis*, G. Pfeffer (**65**), p. 613

Expedition of 1913: Stn. 3439, depth 0–1,000 m. August 26, 42°44′ N, 62°56′ W. Bourée net rapidly retrieved. South of Halifax. Stn. 3452, depth 0–2,500 m. August 28, 43°26′ N, 59°03′ W. Bourée net rapidly retrieved. South of Nova Scotia.

The two specimens taken from these two stations are identical in size, shape and state of preservation. A description of one will apply equally to the other.

This identification is made with certain reservations because these specimens have no tentacles; I therefore know nothing about their characteristic suckers. On the other hand, the relative length of the fin is less in these specimens than that given by Chun. Since he studied only one specimen, the dimensions he indicated are probably not absolute for the species and are perhaps, as elsewhere, variable with age and sex. Due to the present inadequate evidence too much importance should not be attached to this difference.

Minute papillae, corresponding well to those described by Chun, are present on the ventral skin. The shape of the teeth of the brachial suckers is also similar. They resemble those described by Verrill for his *M. agassizi*. But as the latter species is covered with luminous organs, while there are none in *M. cordiformis*, our species under discussion cannot be affiliated to Verrill's species.

The measurements (mm) of these specimens are:

Total length including the ventral arms ..	120
Dorsal length of the body ..	50
Dorsal length of the fin ..	30

The specimens do not exactly correspond to those described by Chun under the name *Mastigoteuthis cordiformis*. However, it can be confirmed that they belong to a very closely affiliated variety of the type species from which they are separated only by differences in size or sex.

Geographic distribution. Sumatra, North America, Atlantic.

94 **Genus *Megalocranchia* G. Pfeffer, 1884**

Megalocranchia hyperborea (Steenstrup), 1856
(Pl. VII, Fig. 7)

1856. *Leachia hyperborea* Steenstrup (**72**), p. 200
1861. *Taonius hyperboreus*, Steenstrup (**73**), p. 83
1881. *Desmoteuthis tenera* Verrill (**83**), p. 412
1885. *Taonius hyperboreus*, Hoyle (**23**), p. 321
1886. *Taonius hyperboreus*, Hoyle (**24**), p. 191
1908. *Desmoteuthis hyperborea*, Pfeffer (**63**), p. 104
1909. *Desmoteuthis hyperborea*, Hoyle (**28**), p. 277
1912. *Megalocranchia hyperborea*, Berry (**3**), p. 1

The very complicated changes required in the synonymy of this species due to confusion and the application of nomenclatural rules were presented by S. Berry in issue No. 932 of *Science* in 1912 (**3***). The facts cited by this author were so convincing that I have decided to accept the name *Megalocranchia* for this genus.

Expedition of 1912: Stn. 3318, depth 0–4,500 m. September 2, 46°14′30″ N, 10°10′30″ W. Bourée net rapidly retrieved. In the open waters of the Gulf of Gascony.

Geographic distribution. To date, the distribution of this species is confined to the North Atlantic. It is exclusively bathypelagic. Greenland, Jan Mayen, Ireland, North Atlantic coast of the United States.

Three specimens were caught. Unfortunately, they are in poor condition since their delicate bodies were crushed, probably from the rapid hauling up of the gear; the folds and cracks became fixed during preservation in formalin.

The measurements (mm) of these specimens are:

	1	2	3
Total length including tentacles	198	205	134

*Corrected to match with the bibliographic citation—General Editor.

Total length without tentacles	188	195	123
Length of first arm (dorsal)	30	32	17
Length of second arm	37	41	21
Length of third arm	51	50	28
Length of fourth arm (ventral)	33	32	19
Length of tentacle	60	65	40
Length of mantle (dorsal)	127	128	80
Length of mantle (ventral)	123	121	76
Length of fin	64	60	34
Width of head (without eyes)	22	23	14

95 The measurements of the width of the head is doubtful because the eyes are missing, making it difficult to precisely define the head. The head is broader at the base than at the apex; the measurement was taken from the base.

The suckers of the third arms suddenly become much larger after the seventh row. This arrangement is maintained for three or four rows, then they become smaller toward the tip. On the arms, the horny ring of the suckers is smooth. On the tentacular club there are four rows of suckers differing very little in size and furnished with a dentate horny ring.

The gladius is strongly enlarged at the level of the fins where, by the folding in of its edges, it forms a smooth cone open for almost its entire length and enclosing the posterior prolongation of the genital gland. The edges meet only at the tip to close the cone along 5 or 6 mm of its length. There is a narrow dorsal ridge along its entire length.

As is known, the funnel is united with the mantle. It is short, ventrally curved and wrinkled inside.

The funnel has a small interesting apparatus which is nothing more than a modification of the gland and of the funnel valve of other cephalopods. The large, slightly sinuous vein, protruding under the skin, descends toward the visceral mass near the funnel. It emerges almost in the middle of the cephalic wall of this organ, then abruptly disappears below the skin. A raised pad is near this point inside the funnel; it is somewhat horseshoe shaped with a basally directed opening. This pad defines an oval-shaped region whose axis is occupied by the terminal portion of the large vein (Pl. VII, Fig. 7). Three rather soft appendages are attached to this pad, one on the median line at the top of the horseshoe and the other two to the right and left, not far from the end of the branches. These small appendages are approximately 4 mm long and less than 1 mm wide; they are slightly flattened which forms a sort of keel along their border. Their end is rounded, and their surface is papillose and granular. They are softer at their tip than at their base and seem to be able to move freely and perhaps vibrate in the water which passes through the funnel. Their position and the nerve endings they contain, their proximity to the brain and otocysts and their relation with the funnel all suggest that they constitute a regulatory organ for respiration and venous circulation. However, it is difficult to define or even give convincing proof of this role.

Cross sections of this apparatus have given the following results: the otocysts are transparent, have thin walls and are large. I could only partially study those of the specimens in bad condition, and I confirmed the presence of two small, club-shaped papillae on the dorsal and ventral termination of the crest separating the two pockets of the organ. These two papillae seem to be soft and non-cartilaginous. They are slightly narrower at their point of insertion than in their free part. This gives them a club-like appearance.

96 *Megalocranchia abyssicola* (Goodrich), 1896

(Pl. XII, Figs. 6–8; Pl. XIV, Figs. 1–4)

1896. *Taonius abyssicola* Goodrich (**21**), p. 17
1900. *Desmoteuthis abyssicola*, G. Pfeffer (**63**), p. 191
1912. *Megalocranchia abyssicola*, S. Berry (**7**), p. 643

Expedition of 1912: Stn. 3222, depth 0–250 m. August 9, 30°45′40″ N, 25°47′ W. Night fishing with Bourée net rapidly retrieved. Locality of Azores.

The label bears the inscription: colorless jelly with traces of black pigment on the body.

This cephalopod is known up to the present only through a very concise description by Goodrich (**21***) who, in 1896, published a summary study on the cephalopods deposited in the Calcutta Museum. This author doubtfully affiliated another much smaller specimen which he has not described to the same species. Moreover, the information pertaining to the larger form is based on external characteristics. It would, therefore, be of interest to revive Goodrich's description and to complete it because it lacks certain points of primary importance.

With regard to *Megalocranchia hyperborea*, I have stated that S. Berry (**7**) classified earlier species belonging to other genera in this genus. *Taonius abyssicola* Goodrich is one of them and, according to Berry's presentation, it should be *Megalocranchia abyssicola* (Goodrich) 1896.

The specimens examined by this author came from the region of the Laccadive Islands, one from a depth of 1,650 m and the other from 1,370 m. The specimen I have studied comes from the Azores. It was caught at night in 250 m water depth with a Bourée net rapidly hauled up. It must therefore be accepted that this animal, which is almost always bathypelagic, makes a nocturnal vertical ascent. It is noteworthy that it has highly developed luminous organs, a fact that was not reported by Goodrich.

With preservation in formalin, the specimen has lost the transparency it had when it was fresh; it has become yellowish-gray and opaque with blackish arms. Chromatophores sparsely scattered on the skin of the mantle can be detected. The eyes are, unfortunately, badly damaged; one has almost completely disappeared and the other has markedly deteriorated. This is unfortunate because of the almost total disappearance of their luminous organs.

The main measurements (mm) of this specimen are:

	Measurement		mm
	Total length including the tentacles	..	125
	Total length without the tentacles	..	116
	Length of tentacles	..	41
	Length of tentacular club	..	14
	Length of the first arm (dorsal)	..	14
97	Length of the second arm	..	20
	Length of the third arm	..	23
	Length of the fourth arm (ventral)	..	21
	Dorsal length of mantle	..	80
	Ventral length of mantle	..	75
	Length of fin	..	24
	Width of fin	..	20

The arms are relatively longer than is normal for this genus where they are very short. From the preceding measurements, the formula 3, 4, 2, 1 can be established to express their order of magnitude, the first dorsal pair is noticeably the smallest. The third pair are more muscular than the others.

The three dorsal pairs are provided with a marginal membrane almost equally well-developed on both edges. These are supported by very large muscular rods issuing from the base of each sucker. The fourth pair of arms has only a single well-developed ventral membrane. The membrane of the dorsal margin is rudimentary and reduced to a small crest supported by slightly protuberant pads. The ventral arms have a very distinct, large keel which seems to be missing in the others.

*Given as [2] in the French original—General Editor.

The membrane uniting the rods is absent in the middle part of the left arm of the second pair. These rods are very long and curled into a crook-shaped hook. The same arrangement, but less marked, is found on the right side. I do not know whether to consider this accidental or connected with hectocotylization.

The arms have two rows of globular suckers along their entire length. The suckers have a short peduncle and are rather closely set. The shape of the globule is not absolutely spherical; it is slightly oval-shaped due to a small swelling that can be observed above the dentate part of the horny ring. The horny ring is circular in shape; Goodrich has given a more or less accurate illustration of it, except that the teeth are lower, more irregular and softer than he indicated. They are so little distinct that they are difficult to see, even more so since they are transparent. They occupy approximately one-third of the horny ring, the remaining two-thirds is smooth.

The tentacles are short; they are almost double the average length of the arms. They are cylindrical and have a clearly demarcated club, with a crook-shaped tip. The club has four longitudinal rows of suckers arranged in oblique series of four. They are very uniformly arranged and the four lines proceed from the base while progressively diminishing up to the extreme tip where only the last three series no longer has four normal
98 suckers, but two and then only one. The suckers in the middle of the club are approximately 1 mm in diameter. These suckers have long peduncles and a horny ring completely ornamented with teeth. The seven or eight larger dark brown, almost black teeth occupy half the circle directed toward the tip of the tentacle; they are very sharply pointed and curve toward the cavity of the sucker. The proximal half of the ring has twelve much smaller and very slightly protruding teeth. The entire ring is formed by a brown plate broadened into a funnel ornamented with fine carvings. The teeth are arranged on the inner edge of the funnel at the entrance to the cavity of the sucker.

A flat area with 18 cups is at the base of the club; these are sessile suckers and probably act as adhesive knobs; they are much smaller than real suckers.

A very reduced membrane supported by rudimentary muscular rods similar to those on the arms is on each side of the tentacular club.

The club has a distinct keel on three-fourths of the side opposite the side with suckers. The keel increases in size toward the tip of the club where it terminates under the last distal suckers.

The stalk of the tentacle is not absolutely cylindrical; it is slightly flattened on the face that leaves the mouth and arrives at the club, determined by two slight crests on the edges of this surface. The latter is beset with two rows of minute sessile suckers so small that they are difficult to detect even under high magnification. Some of the suckers are alternately arranged but most are in pairs which almost touch each other. They are visible along almost the entire length of the tentacle and seem to be absent only at its base.

The funnel is small in relation to the size of the mantle. It is attached to the margin of the mantle. Its tubular outlet is short, curved and has an upturned upper margin.

When one slits open the ventral wall of the funnel, the special organ appears to be a complex thickening in the shape of a soft, irregular plate (Pl. XII, Fig. 6). This organ has been inadequately described by Goodrich. In the specimen I examined, this plate is symmetrical and rises above the point where the large, hitherto deep-seated vein becomes subcutaneous, protuberant and noticeable. This plate is basally bilobed, and on each lobe there is a triangular plate fastened to it by one side with the other sides free and rolled into a cone. The upper half of the plate is bordered by a pad that encloses the two dimples Goodrich mentions. The upper and middle parts of the pad narrow to form a type of free valve which can press on the upper dimple from above and is positioned directly below the tubular outlet of the funnel. In the live animal these pads and valves
99 are close enough to each other to obstruct the outflow of water and slow down or even check the current. They are homologs of the apparatus I have described in *Megalocranchia hyperborea* but their surface area is far greater than that of the three small appendages of the latter species.

The most interesting of the new fact concerning this species is the presence of luminous organs inside the mantle cavity. On vertically slitting the mantle, a complex region is observed between the two gills and apically limited by the anal opening. The mantle extends downward and is enclosed in a transparent membrane (Pl. XIV, Fig. 2).

Above the line joining the base of the two gills there rises a hard protruding quadrangular mass; it is slightly broader at its base than at its apex and has rounded angles. It is divided into two equal, symmetrical parts by a vertical black cord constituting the ink sac on the intestine. The ink duct passes above the quadrangular mass to the anus. It is capped by a foliaceous, membranous plate. The four rounded angles are the luminous organs. Only their surface is seen since they are embedded deep in the mass of surrounding tissue. The two basal organs are much larger than those at the top.

From each side, a curved black line, with its concavity directed toward the median line, unites the basal organ with that of the top; this line distinctly cuts across the silvery white background of the mass of tissues in which they are encased (Pl. XIV, Fig. 3).

The large, lower luminous organ has a transparent, circular, convex cornea through which its dark iridescent contents can be seen. Externally, a black membrane in the shape of a hemispherical pocket marks the base of the organ against the gills. A circular band of white tissue is between the transparent cornea and the ink sac.

The upper organ is much smaller and only reveals its circular, transparent cornea encased in the mass of white tissue. The black contents of the organ are visible through this cornea.

The sagittal section of the entire mass supporting the luminous organs shows that its middle is formed mainly by the liver, but there are black parts on top and especially at the bottom which are due mainly to the ink sac. The luminous organs rest on these black parts. The whitish gland consists of a spongy part, the liver and another smaller surface part which is probably the pancreas.

Due to the great transparency of the mantle and the almost complete absence of chromatophores, the luminous organs have the same luminous effect as they would if they were placed on the external surface of the mantle.

100 The study of cross sections of these organs has provided some interesting results. Since I had unfortunately taken only one luminous organ from each side so as the animal would not be disfigured too greatly, I was only able to obtain a section in one direction. Sections have to be taken along three planes to obtain a comprehensive idea.

Genus *Corynomma* Chun, 1906

Corynomma speculator Chun, 1906

1906. *Corynomma speculator* Chun (**11**), p. 85
1908. *Liguriella podophthalma* Issel (**29**), p. 228
1910. *Corynomma speculator*, Chun (**14**), p. 367
1912. *Corynomma speculator*, G. Pfeffer (**65**), p. 737

Expedition of 1912: Stn. 3170, depth 0–2,500 m. July 21, 38°24′30″ N, 2°24′ E. Large Richard net. South of the Balearic Islands.

A single specimen of this pretty species was captured. It is in excellent condition.

Geographic distribution. South Altantic (between Bahia and Buenos-Aires), Canary Islands, South Equatorial Current of the Indian Ocean.

Genus *Galiteuthis* Joubin, 1898

Galiteuthis armata, Joubin 1898

1898. *Galiteuthis armata* Joubin (**34**), p. 279
1910. *Galiteuthis* (*Taonidium*) *suhmi* Chun (**14**), p. 382

1911. *Galiteuthis phyllura* Berry (**2**), p. 592
1912. *Galiteuthis armata*, G. Pfeffer (**65**), p. 731
1917. *Galiteuthis armata*, Mortara, (**57**), p. 1

Expedition of 1911: Stn. 3064, depth 0–2,700 m. July 21, 38°03′30″ N, 4°46′45″ E. Bourée net rapidly retrieved. Western Mediterranean. Stn. 3084, depth 0–3,000 m. July 28–29, 34°20′40″ N to 33°46′ N, 10°05′ W to 10°44′ W. Small Bourée net. Between Gibraltar and the Salvages Islands. Stn. 3147, depth 0–550 m. August 26, 38°12′ N, 25°49′ W. Bourée net rapidly retrieved. Azores.

Expedition of 1912: Stn. 3202, depth 0–3,000 m. August 5, 31°45′ N, 20°17′ W, Bourée net rapidly retrieved. Near Madeira. Stn. 3281, depth 0–4,000 m. 39°25′40″ N, 35°14′30″ W. Bourée net rapidly retrieved. Azores.

Expedition of 1913: Stn. 3452, depth 0–2,500* m. August 28, 43°26′ N, 59°03′ W. Bourée net rapidly retrieved. Southwest of Nova Scotia.

Specimen from Stn. 3064. This is a rather large individual with a total length of 291 mm in rather bad condition. The fin is 105 mm long, 40 mm of which constitute its filiform tip. The eyes are large, very protuberant
101 and have a large, luminous plaque in a clear crescent form. The distance from one crystalline lens to the other is 24 mm. The eyes are large enough to almost completely screen the head so that only a narrow vertical band can barely be seen between them.

Specimen from Stn. 3068. Length 188 mm including the tentacles. Its fin is leaf-shaped, well formed and terminates in a point whose length is equal to approximately one-third of the entire fin. The two lenses are 15 mm apart. Each eye is approximately as large as the head. A crescent-shaped luminous organ is on the globe below each eye, the upper concavity bordering the lens.

Specimen from Stn. 3084. Total length of the 1st specimen is approximately 80 mm; it is in very bad condition. The fin is prolonged by a large filiform appendix.

The second specimen is very large with a total length of 610 mm. This must be the largest described to date. The fin is 180 mm long and terminates in a very short point only about 30 mm long. The eyes are enormous; unfortunately, only one remains attached to the head. The tentacles are 185 mm long.

Specimen from Stn. 3147. This is a small individual, 110 mm in total length, in bad condition. The tentacular club is missing. The leaf-shaped fin is short and terminates in a small point; it is 17 mm long, 5 mm of which constitutes the caudal tip. The lenses are approximately 10 mm apart.

Specimen from Stn. 3202. A small specimen with a total length of 75 mm. It has lost its tentacles; the tail is very fragile; the fin is poorly developed.

Specimen from Stn. 3281. Strongly resembles the specimen from Stn. 3068. Total length is 180 mm. The specimen is in rather bad condition. The caudal tip is equal to approximately one-third the length of the fin. The two lenses are 13 mm apart. The luminous organs are arranged as mentioned above, but they are smaller. The eyes are smaller than the head.

Specimen from Stn. 3452. This is a large individual with a total length of 480 mm including the tentacles. The fin is 145 mm long, 35 mm of which constitute the posterior end; it is less fragile and shorter than in the other specimens. Only one wide eye is left which appears to have been very large and having a flat luminous organ, but so deteriorated that it is impossible to precisely define the contour and to compare it to the other specimen.

Measurements taken of various specimens indicate that the proportion of the eyes to the width of the head increases with age. They are relatively larger in adults than in juveniles.

It is also noteworthy that in very young individuals, the fin is prolonged by a short filiform appendix.
102 This appendix is much longer in individuals of moderate size and becomes equal to one-third the total length

*French original states 25,000 m, a typographical error—Scientific Editor.

of the fin. On the contrary, it is short in the very large individuals and represents scarcely one-fifth the length of the fin. Modifications have probably taken place due to the development of the reproductive organs. In all cases they show that, in this cephalopod as in many others, the measurements given to describe a species can only be given a very secondary value since they can vary in wide proportions with the age and sexual maturity of the individuals. It is not possible to formulate a concise rule for *Galiteuthis armata* without having studied a large number of individuals in good condition and whose sex is verified.

Geographic distribution. Mediterranean, Messina, Guinea Current, California, Azores, Nova Scotia.

Family CRANCHIADAE

Gen. sp. ?

Expedition 1911: Stn. 3079, surface. July 27, 35°06′ N, 8°25′ W. Fine-meshed net. Near Gibraltar.

A very young specimen in bad condition, approximately 5 mm long and completely contracted in alcohol. The mantle seems to be united to the base of the funnel. The arrangement of the rudimentary arms suggests that this specimen belongs to a genus of Cranchiidae but it is impossible to give more precise information.

Stn. 3089, depth 0–4,000 m. July 30, 32°21′30″ N, 12°31′ W. Large Richard net. Between Gibraltar and the Ilhéus Selvagens.

A very young highly contracted specimen 4 to 5 mm long, impossible to identify.

Tables of Species Collected from Different Stations

EXPEDITION OF 1910

Station No.	Date	Locality		Depth in meters	Nature of bottom	Method of collection	Species collected
		Latitude	Longitude (Greenwich)				
3015	30 Aug.	41°02′ N	10°20′ W	0–4,200		Bourée net rapidly retrieved	*Mastigoteuthis grimaldii*
3028	4 Sept.	36°54′30″ N	11°49′ W	0–5,100		”	*Mastigoteuthis grimaldii*
3036	7 Sept.	36°06′40″ N	10°18′ W	0–4,740		”	*Mastigoteuthis grimaldii*
3053	13 Sept.	37°20′ N	0°07′ W		Surface	Fine-meshed net	*Rhynchoteuthion*

EXPEDITION OF 1911

Station No.	Date	Locality: Latitude	Locality: Longitude (Greenwich)	Depth in meters	Nature of bottom	Method of collection	Species collected
3064	21 July	38°03′30″ N	4°46′45″ E	0–2,700		Bourée net rapidly retrieved	*Abralia morrisii*, *Calliteuthis* sp. juv., *Galiteuthis armata*
3068	22 July	37°16′ N	1°15′ E	0–2,600		,,	*Galiteuthis armata*
3078	27 July	35°18′15″ N	7°52′ W	0–2,000		,,	*Mastigoteuthis grimaldii*
3079	27 July	35°06′ N	8°25′ W	Surface		Fine-meshed net	Cranchiidae
3082	28 July	34°20′40″ N	10°05′ W	0–4,000		Bourée net rapidly retrieved	*Bolitaena (Eledonella) diaphana*
3084	28–29 July	34°20′40″ N 33°46′ N	10°05′ W 10°44′ W	0–3,000		,,	*Chiroteuthis picteti*, *Galiteuthis armata*
3086	29 July	33°46′ N	10°44′ W	0–4,300		,,	*Bolitaena (Eledonella) diaphana*, *Chiroteuthis picteti*, *Mastigoteuthis grimaldii*
3089	30 July	32°21′30″ N	12°31′ W	0–4,000		Large Richard net	*Pyroteuthis* sp. juv., Cranchiidae
3097	1 Aug.	30°03′ N	15°56′ W	Surface		Fine-meshed net	*Rhynchoteuthion*
3100	2 Aug.	29°03′ N	16°08′30″ W	0–3,500		Bourée net rapidly retrieved	*Chiroteuthis portieri*
3122	12 Aug.	Southern point of the Desert Grande Island		Surface		Caught at low tide	*Octopus vulgaris*
3126	14 Aug.	32°34′45″ N	17°05′30″ W	Surface		Eel-speared under electric light	*Stenoteuthis bartrami*, *S. pteropus*
3128	15 Aug.	33°40′ N 33°52′ N	19° W 19°16′ W	0–3,500		Three-layered net	*Bolitaena (Eledonella) diaphana*, *Vitreledonella alberti*
3129	15 Aug.	34°05′ N	19°37′ W	Surface		Fine-meshed net	*Thelidioteuthis alessandrinii*
3131	16 Aug.	35°09′ N	21°21′ W	0–3,500		Bourée net rapidly retrieved	*Bolitaena (Eledonella) diaphana*, *Vitreledonella alberti*, *Abralia morrisii*, *Mastigoteuthis grimaldii*, *Benthoteuthis megalops*
3147	26 Aug.	38°12′ N	25°49′ W	0–550		Bourée net rapidly retrieved	*Pyroteuthis margaritifera* var. *aurantiaca*, *Brachioteuthis beani**, *Galiteuthis armata*
3150	27 Aug.	38°01′ N	25°21′ W	1,740		Trawl	*Pyroteuthis margaritifera* var. *aurantiaca*

*This species, illustrated in Pl. XIV, Figs. 5 and 6, will be described later in the text.

EXPEDITION OF 1912

Station No.	Date	Locality: Latitude	Locality: Longitude (Greenwich)	Depth in meters	Nature of bottom	Method of collection	Species collected
3170	21 July	38°24′30″ N	2°24′ E	0–2,500		Large Richard net	*Corynomma speculator*
3190	27 July	34°27′ N	14°15′ W	0–3,000		Bourée net rapidly retrieved	*Thelidioteuthis alessandrinii*
3202	5 Aug.	31°45′ N	20°17′ W	0–3,000		,,	*Abraliopsis pfefferi, Abralia morrisii, Benthoteuthis megalops, Tracheloteuthis riisei, Galiteuthis armata*
3205	6 Aug.	31°01′ N	23°45′ W	0–3,000		,,	*Bolitaena (Eledonella) diaphana*
3207	6 Aug.	30°54′ N	24°11′ W	Surface		Shrimp-net under electric light	*Onychoteuthis banksi, Stenoteuthis bartrami*
3213	7 Aug.	30°45′40″ N	25°47′ W	0–500		Bourée net rapidly retrieved	*Cycloteuthis sirventi, Ommatostrephes sagittatus*
3215	8 Aug.	30°45′30″ N	25°47′ W	0–2,000		,,	*Bolitaena (Eledonella) diaphana*
3217	8 Aug.	30°45′30″ N	25°47′ W	0–1,000		,,	*Bolitaena (Eledonella) diaphana, Abraliopsis pfefferi, Pyroteuthis margaritifera, Chiroteuthis veranyi, Calliteuthis meneghinii*
3222	9 Aug.	30°45′40″ N	25°47′ W	0–250		Bourée net rapidly retrieved	*Megalocranchia abyssicola*
3223	10 Aug.	30°50′ N	25°43′ W	5,300		Trawl	*Vitreledonella richardi*
3256	19 Aug.	38°16′40″ N	28°23′ W	0–1,200		Bourée net rapidly retrieved	*Heteroteuthis dispar*
3279	23 Aug.	38°55′ N	34°07′30″ W	0–3,000		,,	*Pyroteuthis margaritifera, Histioteuthis bonelliana, Ommatostrephes sagittatus, Calliteuthis meneghinii*
3281	24 Aug.	39°25′40″ N	35°14′30″ W	0–4,000		,,	*Pyroteuthis margaritifera, Mastigoteuthis grimaldii, Galiteuthis armata, Histioteuthi bonelliana, Calliteuthis meneghinii*

EXPEDITION OF 1912 (*Contd.*)

Station No.	Date	Locality		Depth in meters	Nature of bottom	Method of collection	Species collected
		Latitude	Longitude (Greenwich)				
3284	24 Aug.	39°19′ N	35°24′30″ W	0–1,000		Bourée net rapidly retrieved	*Pyroteuthis margaritifera, Histioteuthis bonelliana, Ommatostrephes sagittatus*
3285	24 Aug.	39°23′ N	35°18′30″ W	0–500		,,	*Meleagroteuthis hoylei, Calliteuthis meneghinii, Histioteuthis bonelliana*
3286	25 Aug.	39°19′ N	35°40′ W	0–1,000		,,	*Onychoteuthis banksi*
3312	1 Sept.	45°02′ N	13°05′ W	0–3,500		,,	*Calliteuthis meneghinii, Calliteuthis?*
3318	2 Sept.	46°14′30″ N	10°10′30″ W	0–4,500		,,	*Megalocranchia hyperborea, Gonatus fabricii*
3340	17–30 Sept.	En route to Thamshavn (Norway)		5–15		Trammel net	*Rossia palpebrosa*

EXPEDITION OF 1913

Station No.	Date	Locality		Depth in meters	Nature of bottom	Method of collection	Species collected
		Latitude	Longitude (Greenwich)				
3360	24 July	47°06′30″ N	8°11′ W	0–3,500		Bourée net rapidly retrieved	*Stigmatoteuthis verrilli*
3383	4 Aug.	38°40′ N	34°15′ W	Surface		Fine-meshed net	*Rhynchoteuthion*
3412	9 Aug.	39°59′ N	56°20′ W	5,270		Trawl	*Cirroteuthis umbellata*, *Stigmatoteuthis verrilli*
3414	10 Aug.	40°15′ N	56°25′ W	0–4,000		Bourée net rapidly retrieved	Enoploteuthide?, *Stigmatoteuthis verrilli*
3415	10 Aug.	40°15′ N	56°17′ W	0–2,000		,,	*Mastigoteuthis grimaldii*
3437	26 Aug.	42°40′ N	62°49′30″ W	1,458		Trawl	*Graneledone verrucosa*
3439	26 Aug.	42°44′ N	62°56′ W	0–1,000		Bourée net rapidly retrieved	*Chiroteuthis lacertosa*, *Mastigoteuthis cordiformis*, *Tracheloteuthis riisei*
3447	27 Aug.	43°18′ N	60°11′ W	0–1,100		,,	*Stigmatoteuthis verrilli*
3452	28 Aug.	43°26′ N	59°03′ W	0–2,500		,,	*Stigmatoteuthis verrilli*, *Mastigoteuthis grimaldii*, *M. cordiformis*, *Galiteuthis armata*, *Illex illecebrosus*
3453	28 Aug.	43°26′ N	59°03′ W	0–1,000		,,	*Stigmatoteuthis verrilli*, *Gonatus fabricii*, *Tracheloteuthis riisei*
3473	6 Sept.	42°36′30″ N	63°36′30″ W	1,332		Trawl	*Rossia caroli*, *Stigmatoteuthis verrilli*, *Calliteuthis meneghinii*
3526	29 Sept.	38°30′ N	34°40′ W	0–2,000		Large Richard net	*Abraliopsis* sp. juv., *Mastigoteuthis grimaldii*

EXPEDITION OF 1914

Station No.	Date	Locality				Depth in meters	Nature of bottom	Method of collection	Species collected
		Latitude		Longitude (Greenwich)					
3591	26 July	42°18′	N	12°08′	W	0–3,500		Bourée net, closing type	*Calliteuthis meneghinii*

Bibliography

1. Appellöf, A. 1890. Teuthologische Beiträge. I. *Ctenopteryx* n. g., *Verania sicula* Krohn, *Calliteuthis* Verrill. Bergens Museums Aarsberetning for 1889. Bergen.
2. Berry, S. 1911. Preliminary notices of some new Pacific Cephalopods. *Proc. U. S. National Museum*, Vol. 11, Washington.
3. Berry, S. 1912. Some necessary changes in Cephalopod nomenclature. *Science*, N. S., Vol. 36, No. 932, p. 643, Nov. New York.
4. Berry, S. 1910. A review of the Cephalopods of western north America. *Bull. Bureau of fisheries*, Vol. XXX, Washington (published July 24, 1912).
5. Berry, S. 1916. Cephalopods of the Kermadec Islands. *Proc. Academy Natural Sciences of Philadelphia*, Vol. 68, p. 45–69, March, Philadelphia.
6. Berry, S. 1917. Australian antarctic expedition 1911–1914. Cephalopoda, Series C. *Zoology and Botany*, Vol. 4, p. 2. 31 March, Adelaide.
7. Berry, S. 1917. On *Moschites verrucosa* (Verrill) and its allies. *Proc. Acad. Nat. Sc. Philadelphia*, Vol. 69, January, Philadelphia.
8. Carus, V. 1889–1893. Prodromus Faunae Mediterraneae. Cephalopoda: II, Part II, Stuttgart.
9. Chun, C. 1903. *Rhynchoteuthis*. Eine merkwürdige Jugendform von Cephalopoden. *Zool. Anz.*, XXVI, Leipzig.
10. Chun, C. 1903. Über Leuchtorgane und Augen von Tiefsee-Cephalopoden, *Verh. Deutsche Zool. Ges.* XIII, Berlin.
11. Chun, Carl. 1906. System der Cranchien. *Zoolog. Anzeiger.*, Leipzig.
12. Chun, Carl. 1908. Über Cephalopoden der deutsche Tiefsee-Expedition. *Zool. Anzeiger*, XXXIII, Leipzig.
13. Chun, Carl. 1910. *Cirrothauma* ein blinder Cephalopod. Leipzig.
14. Chun, Carl. 1910. Wissenschaftliche Ergebnisse der deutschen Tiefsee-Expedition. Die Cephalopoden, Teil I. Oegopsida. Leipzig.
15. Cuénot, L. 1914. Les organes phagocytaires des Mollusques. *Arch. Zool. Expér.*, Vol. 54, No. 9, p. 267, July, Paris.
16. Fabricius, O. 1780. Fauna Groenlandiae, etc. Hafniae et Lipsiae.
17. Férussac, J.B.L. 1835. Note sur deux genres de Céphalopodes encore peu connus, les genres *Calmaret* (*Loligopsis*) et *Cranchia. Magasin de zoologie* I. Paris.
18. Férussac, J.B.L. 1835–1848. Histoire naturelle générale et particulière des Mollusques, etc. Paris.
19. Fischer, Paul. 1883. Note préliminaire sur une nouvelle espece du genre *Cirroteuthis*. *Journ. de. Conch.* Vol. 31. Paris.
20. Fischer, H. and L. Joubin. 1907. Expéditions scientifiques du *Travailleur* et du *Talisman*. Céphalopodes. Vol. VIII. Paris.
21. Goodrich, E.S. 1896. Report on a collection of Cephalopods from the Calcutta Museum. *Trans. Linn. Soc.*, 2nd series, Vol. VII, part I, December, London.
22. Gray, J.E. 1849. Catalogue of the Mollusca in the collection of the British Museum, part I. Cephalopoda antepedia, London.

23. Hoyle, W.E. 1885. Diagnoses of new species of Cephalopoda collected during the cruise of *H. M. S. Challenger*. I. The Octopoda. *Ann. and Mag. of Nat. Hist.* [V] Vol.15, London—II. The Decapoda, Ibid, [V] Vol. 16.
24. Hoyle, W.E. 1886. Report on the scientific results of the voyage of *H.M.S. Challenger* during the years 1873–1876. *Zoology*, Vol. 16, part XLIV, Report on the Cephalopoda. Edinburgh.
25. Hoyle, W.E. 1889. Observations on the anatomy of a rare Cephalopod (*Gonatus fabricii*). *Proc. Zool. Soc.*, p. 117, London.
26. Hoyle, W.E. 1902. The luminous organs of *Pterygioteuthis margaritifera*, a mediterranean Cephalopod. *Mem. Manchester Soc.*, XLVI, No. 16, Manchester.
27. Hoyle, W.E. 1904. Reports on the dredging operations, carried on by the *Albatross*. *Bull. Mus. Compar. Zool.* Vol. 43, Cambridge.
28. Hoyle, W.E. 1886. A catalogue of recent Cephalopods. *Proc. Roy. Phys. Soc.*, Vol. 9, Edinburgh; 1st Supplement 1897—2nd Supplement 1906.
29. Issel, Raffaele. 1908. Raccolte planctoniche fatte dalla R. Nave *Liguria* nel viaggio di circonnavigazione del 1903–1905. IV. Molluschi, Parte I. Cefalopodi planctonici. Firenze.
30. Jatta, Guiseppe. 1896. Cefalopodi viventi nel golfo di Napoli, in: Fauna und Flora des Golfes von Neapel Monog, 23: Cefalopodi (Systematica), Berlin.
31. Joubin, Louis. 1894. Céphalopodes d'Amboine. *Revue Suisse de Zoologie II*, Genève.
32. Joubin, Louis. 1895. Contribution à l'étude des Céphalopodes de l'Atlantique Nord. Rés. Camp. Scient. accomplies sur son yacht par S.A.S. Albertb Ier, Prince Souverain de Monaco. Fasc. IX, Monaco.
33. Joubin, Louis. 1896. Observations sur divers Céphalopodes. Note I: *Abraliopsis pfefferi* nov. gen. et sp. *Bulletin Soc. Scient. et Médic. de l'Ouest*, Vol. 5, No. 1, p. 19–35.
34. Joubin, Louis. 1898. Note sur une nouvelle famille de Céphalopodes. *Ann. des. Sc. nat. Zoologie*, VI, Paris.
35. Joubin, Louis. 1900. Céphalopodes provenant des Campagnes de la *Princesse-Alice* (1891–1897). Résultats des campagnes scientifiques accomplies sur son yacht par S.A.S. Albert Ier, Prince Souverain de Monaco. Fasc. XVII, avec quinze planches. Monaco.
36. Joubin, Louis. 1902. Observations sur divers Céphalopodes. Sixième note: Sur une nouvelle espèce du genre *Rossia*. *Bulletin de la Société Zoologique de France*, Vol. 27, p. 138–143, Paris.
37. Joubin, Louis. 1902. Révision des Sepiolidae. *Mém. Soc. Zool. de France*, XV, p. 80 to 145.
38. Joubin, Louis. 1903. Sur quelques Céphalopodes recueillis pendant les dernières campagnes de S.A.S. le Prince de Monaco (1901–1902). *Comptes rendus de l'Académie des Sciences*, Vol. 136, p. 100–102, Paris.
39. Joubin, Louis. 1905. Note sur les organes lumineux de deux Céphalopodes. *Bull. Soc. Zool. de France*, XXX, Paris.
40. Joubin, Louis. 1912. Note sur les Céphalopodes capturés en 1911 par S. A. S. le Prince de Monaco. *Comptes rendus de l'Académie des Sciences*, 154, février, Paris.
41. Joubin, Louis. 1920. Céphalopodes provenant des Campagnes de la *Princesse-Alice* (1898–1910), (3e Série) Résultats des Camp. scient. accomplies sur son yacht par S. A. S. Albert Ier, Prince Souverin de Monaco, etc. Monaco.
42. Joubin, Louis. 1916. Études préliminaires sur les Céphalopodes recueillis au cours des croisières de S.A.S. le Prince de Monaco. 4e Note: *Chiroteuthis portieri*. *Bull. Inst. Océanographique*, No. 317, Monaco.
43. Joubin, L. 1918. Études préliminaires sur les Céphalopodes recueillis au cours des croisières de S. A. S. le Prince de Monaco. 5e Note: *Moschites verrucosa* (Verrill). *Bull. Inst. Océanographique*, No. 339. Février, Monaco.
44. Joubin, L. 1918. Études préliminaires sur les Céphalopodes recueillis au cours des croisières de S. A. S. le Prince de Monaco. 6e Note: *Vitreledonella richardi* Joubin. *Bull. Inst. Océanographique*, No. 340, Monaco.

45. Joubin, L. 1919. Études préliminaires sur les Céphalopodes recueillis au cours des croisières de S. A. S. le Prince de Monaco. 7e Note: *Cycloteuthis sirventi* nov. gen. et sp. *Bull. Inst. Océanographique*, No. 351, Avril, Monaco.
46. Joubin, L. 1920. Études préliminaires sur les Céphalopodes recueillis au cours des croisières de S. A. S. le Prince de Monaco. 8e Note: Quelques points de l'Anatomie d'un Céphalopode abyssal, *Eledonella diaphana* Hoyle. *Bull. Inst. Océanographique*, No. 382, December 30, Monaco.
47. Lamarck, Le Citoyen. 1799. Sur les genres de la Seche, du Calmar et du Poupe, vulgairement nommés Polypes de mer. *Mém. Soc. Hist. Nat.*, Vol. 1, Paris.
48. Lamarck, Le Chevalier de. 1814–1822. Histoire naturelle des Animaux sans vertèbres, présentant les caractères généraux et particuliers de ces animaux, leur distribution, leurs classes, leurs familles, leurs genres, et la citation des principales espèces qui s'y rapportent. Vol. 6, 1814, Paris; Vol. 9, 1822, Paris.
49. Leach, W.E. 1817. Zoological miscellany, being description of new or interesting animals, illustrated with colored figures drawn from nature by R.P. Nodder. vol. 3, London.
50. Leach, W.E. 1852. Molluscorum Britanniae synopsis. A synopsis of the Mollusca of Great Britain, arranged according to their natural affinities and anatomical structure. London.
51. Lesueur. 1821. Description de plusieurs espèces nouvelles de Seiches. *Journ. of the Nat. Acad. of Science*, Vol. 2, Philadelphia.
52. Lichtenstein, H. 1818–1918. Von den Sepien mit Krallen. *Abhdl. Berlin Acad.*, Vol. 7, Berlin.
53. Lo Bianco. 1903. Le pesche abissali eseguiti da F.A. Krupp col yacht *Puritan* nelle adiacenze de Capri. *Mitt. Staz. Neapel*, Berlin, XVI, p. 109, Berlin.
54. Lönnberg, E. 1898. On the cephalopods collected by the Swedish Expedition to Tierra del Fuego. *Svenska Exped. Magellansländern II*, p. 49–64, Stockholm.
55. Massy, A.L. 1907. Preliminary notice of new and remarkable Cephalopods from the South West Coast of Ireland. *Ann. and Mag. Nat. Hist.*, [7] XX, p. 377, London.
56. Möller. 1842. Index molluscorum Groenlandiae. *Kröyer Nat. Hist. Tidsskr*, IV, Copenhague.
57. Mortara, Silva. 1917. *Galiteuthis armata* Joubin e *Galiteuthis phillura*, Berry R. Comitato talassografico italiano, Memoria LX Venezia.
58. Nishikawa, T. 1906. On a rare Cephalopod. *Zool. Mag. T.* XVIII, Tokyo.
59. Norman, A.M. 1890. Revision of British Mollusca. *Ann. and Mag. of Nat. Hist.*, [VI], Vol. 5, London.
60. Orbigny, A. d' and Férussac, D. de. 1835–1848. Histoire naturelle générale et particulière des Céphalopodes acétabulifères vivants et fossils. Paris.
60b. Owen, R. 1881. Description of some new and rare Cephalopoda. *Trans. Zool. Soc.*, Vol. 11, London.
61. Pfeffer, G. 1884. Die Cephalopoden des Hamburger naturhistorischen Museums. *Abhandl. aus dem Geb. der Nat. heraus v. naturwissensch. Verein*, Vol. 8, Hamburg.
62. Pfeffer, G. 1900. Synopsis der oegopsiden Cephalopoden. *Mittheil aus dem Naturhistorischen Museum*, XVII, Hamburg.
63. Pfeffer, G. 1908. Teuthologische Bemerkungen. *Mitteil. Naturhistorische Museum*, XXV, Jahr. Hamburg Wiss. Anstalt, XXV, p. 187–295, Hamburg.
64. Pfeffer, G. 1908. Cephalopoden, in Brandt und Apstein Nordisches Plankton, IX, p. 9–116. Kiel.
65. Pfeffer, G. 1912. Die Cephalopoden der Plankton Expedition, Ergebnisse der Plankton-Expedition der Humboldt Stiftung. Kiel and Leipzig.
66. Prosch, V. 1849. Nogle nye Cephalopoder beskrevne og anatomisk undersögt. *Overs. Danske vid. Selskabs. Forh. I.*, Kjøbenhavn.
67. Rochebrune, T. de. 1884. Etude Monographique de la famille des Loligopsidae. *Bull. de la Soc. Philomatique*, [VII], Vol. 8, Paris.
68. Rüppell, E. 1845. Intorno ad alcuni Cefalopodi del mare di Messina. *Giorn. del gabinetto di Messina*, XXVII.

69. Russel, E.S. 1909. Preliminary notice on the Cephalopoda collected by the Fishery cruiser *Goldsecker* 1903–1908. *Ann. and Mag. Nat. Hist.* [8], III, p. 446, London.
70. Sasaki, M. 1916. Notes on Oegopsid Cephalopods found in Japan. *Annot. Zool. Jap.*, Vol. IX, Tokyo.
71. Steenstrup, J. 1855. Kjäber af en Kolossal Blacksprutte. *Overs. K. vid. Selbsk. Forh.*, p. 199, Kjøbenhavn.
72. Steenstrup, J. 1856. Hectocotydannelsen. *Vidensk. Selsk. Skr.* [5], Vol. 4, p. 185, Kjøbenhavn.
73. Steenstrup, J. 1861. Cephalopodum, qui in museis Hafniensibus inveniuntur, Pelagicorum synopseos particula prima. *Overs. Dansk. videns Selsk.* Kjøbenhavn.
74. Steenstrup, J. 1880. De Ommastrephagtige Bläcksprutters indbyrdes Forhold. *Overs. Dansk. Vid. Selsk. Forh.* Kjøbenhavn.
75. Steenstrup, J. 1881. Professor Verrill's to nye Cephalopodslägter *Stenoteuthis* og *Lestoteuthis*. *Overs. K. dansk. Videnskab. Selsk. Forhand.* p. 1, Kjøbenhavn.
76. Steenstrup, J. 1881. En ny Blaecksprutterslaegt: *Tracheloteuthis*. *Vidensk. meddel. Natur. Forh.* Kjøbenhavn.
77. Troschel, H. 1857. Bemerkungen über die Cephalopoden von Messina. *Arch. f. Naturges*, 23.
78. Vérany, J.B. 1837. Mémoires sur deux nouvelles espèces de Céphalopodes trouvés dans l'Océan. *Mem. Accad. Torino*, p. 99, Torino.
79. Vérany, J.B. 1839. Mémoire sur six nouvelles espèces de Céphalopodes trouves dans la Méditerranée. *Mem. R. Accad. d. Sc.* [II], Vol. 2, Torino.
80. Vérany, J.B. 1851. Mollusques méditerranéens observés, décrits, figurés et chromolithographiés d'aprés nature sur des modèles vivants, 1er part, Céphalopodes de la Méditerranée. Gênes.
81. Verrill, A.E. 1880. Synopsis of the Cephalopoda of the North Eastern coast of America. *Amer. Journ. Sc. and Arts*, Vol. 20, New Haven.
82. Verrill, A.E. 1881. Report on the Cephalopods and on some additional species dredged by the U. S. Fish commission steamer *Fishhawk* during the season of 1880. *Bull. Mus. Comp. Zoöl. at Harvard College*, Vol. 8, No. 5, Cambridge.
83. Verrill, A.E. 1881. The Cephalopods of the North Eastern coast of America. Part II: The smaller Cephalopods including the Squids and the Octopi with other allied forms. *Trans. Connecticut Academy*, Vol. 5, New Haven.
84. Verrill, A.E. 1882. Report on the Cephalopods of the North Eastern coast of America. Annual report of the commissioner of Fish and Fisheries for 1879. Washington.
85. Verrill, A.E. 1881–1885. Second catalogue of Mollusca recently added to the fauna of the New England coast and the adjacent parts of the Altantic. consisting mostly of deep sea species, with notes on others previously recorded. *Trans. Connect. Acad.*, Vol. 6, New Haven.
86. Verrill, A.E. 1885. Third catalogue of Mollusca recently added to the fauna of the New England coast and the adjacent parts of the Atlantic. *Trans. Connect. Acad.*, Vol. 7, New Haven.

Plates

PLATE I

Figs. 1 and 2. *Vitreledonella richardi* Joubin

Fig. 1. Dorsal side. Water-color sketch by Miss Vesque of a specimen preserved in sea water and formalin. Natural size.

Fig. 2. Ventral side.

PLATE II

PLATE IV

Figs. 1 to 9. *Vitreledonella richardi* Joubin

Fig. 1. Web between the two ventral arms. Natural size.

Fig. 2. View of the superior orifice of a sucker. The nerve ganglion is on the left. Approx. × 25.

Fig. 3. Side view of a sucker, showing its innervation and its position in the hyaline nipple. Below this point the small nerve ganglion is seen on the track of the nerve which lies at the base of the sucker at the ganglion of the branchial nerve. Approx. × 25.

Fig. 4. Dorsal view of the brain; showing 1, 2, 3 and 4 as brachial nerves; (B) buccal ganglion; (O) esophagus; (P) pallial nerve; (R) optic nerve; (S) salivary duct; (U) commissure going to the buccal ganglion; (V) vein; (X) olfactory nerve; (Z) ophthalmic nerve. × 15.

Fig. 5. Front view of the eye. Approx. × 5.

Fig. 6. Optic ganglion; (C) predominant ganglion; (M) oval spot; (P) optic nerve. × 35.

Fig. 7. Upper jaw, viewed from three-fourths. × 10.

Fig. 8. Lower jaw, viewed from above. × 10.

Fig. 9. Ovary cut open to show the thick hump-like gland (R) in the interior; (I) intestine; (J) oviduct. The outer membrane with violet chromatophores was dissected as well as folded back on right side; in the bottom of the pocket, the chromatophores are visible owing to transparency. Approx. × 5.

PLATE V

Figs. 1 to 20. *Vitreledonella richardi* Joubin

Fig. 1. Schematic arrangement of suckers on an arm. Reduced to one-third.

Fig. 2. Section passing through the brachial ganglion (A) artery; (C) cortical cells; (E*) axial nerve fiber; (G) upper large cells; (T) myeline fibers; (V) vein. × 150.

Fig. 3. Section passing through brachial nerve between two ganglia. (A) artery; (V) vein. × 150.

Fig. 4. Section through the thin middle part of a sucker. (C) circular muscle fibers; (E) internal epithelial layer; (O) intercrossed layer of muscle fibers; (P) bundle of the transverse fibers; (M) superficial muscle layer. × 80.

Fig. 5. Section in the region of the lip of a sucker. (E) internal epithelium; (L) bundle of longitudinal muscle fibers. × 80.

Fig. 6. General disposition of central nervous system.
1, 2, 3, 4—brachial nerves of four arms; (B) buccal ganglion; (C) cerebral ganglion; (D) white body on ocular globe; (F) fibers of the rudimentary capsule in which the central nervous system is suspended; (G) optic ganglion; (M) white body; (N) pallial nerve; (O) esophagus; (P) pedal ganglion; (S) salivary canal; (U) visceral ganglion; (V) vein; (W) peduncular ganglion. Approx. × 4.

Fig. 7. Branching of the optic nerve on the eyeball. (A) inferior branch of the optic nerve; (B) white body; (C) peduncular ganglion; (D) superior branch of the optic nerve; (P) optic nerve; (N) terminal swellings of the optic nerve. × 30.

Fig. 8. Section of retina. The right part (X) of the figure shows the club-like swellings of the rod cells in the folded regions.
(A) region of nuclei; (B) rods; (C) layer of fibrous capsule; (D) padding of the optic nerve; (E) connective envelope; (L) external limitation; (M) section of the white body; (P) section of retinal pigment layer; (S) inner limitation; (X*) club-like rods. × 360.

Fig. 9. General arrangement of visceral organs.
(A) anus; (B) gill; (E*) stomach; (G) genital gland; (L) oviducal gland; (N) brain; (O) genital aperture; (R) renal aperture; (S) pallial ganglion on the pallial bridle; (I) ink sac. Approximately natural size.

Fig. 10. Digestive system. (B) buccal bulb; (C) brain; (F**) liver (G, H) anal ampulla with the opening of the ink duct; (I) intestine; (J) a string of ink (a rosary of ink beads); (K) genital gland; (L) crop; (M) ink gland; (N) ink reservoir; (O) esophagus. Approx. × 5.

Fig. 11. Side view of the buccal bulb. (B) buccal ganglion; (G) pharyngeal ganglion; (O) esophagus; (N) cerebro-pharyngeal commissures; (P) pharyngeal appendage. Approx. × 10.

Fig. 12. Half of the row of radular teeth. × 20.

Fig. 13. Central part of the digestive system. The membrane with chromatophores

*Not indicated in the figure.

**Not indicated in the illustration—General Editor.

and golden iridocytes has been held back to one side. (A) gastric ganglion; (F) liver; (G) salivary gland; (H) hepatic duct; (I) intestine; (K) genital gland; (L) stomach; (N) top of the ink gland; (W) spiral stomach or caecum; (Y) crop. Approx. × 15.

Fig. 14. Spiral caecum and its adjoining organs. (F) liver; (G) gastric ganglion; (H) hepatic duct with considerable swelling; (I) intestine; (M) terminal part of the stomach; (L) stomach; (S) caecum. Approx. × 15.

Fig. 15. Left gill, enlarged approximately five times. The upper figure 15^{I} shows the lower convex side and the lower figure 15^{II} shows the upper concave side. (A) efferent artery; (G) branchial gland; (N) branchial ganglion and nerves; (V) afferent vein. Approx. × 5.

Fig. 16. Schematic diagram of arrangement of vessels, nerves, glands and respiratory tissue. (A) afferent artery; (N) nerve; (G) branchial gland; (V) afferent vein.

Fig. 17. Heart on the ventral side.
(A) aorta; (C) genital artery; (E) hepatic artery(?); (V) renal artery; (D) ventricle; (H) auricle. Approx. × 5.

Fig. 18. Disposition of the excretory system.
(A) aorta; (D) kidney; (N) vein (O) genital aperture; (R) renal aperture; (S) venous (branchial) heart; (T) oviducal gland; (U) urinary sac; (V) branchial efferent artery; (Y) branchial afferent vein from venous heart. Approx. × 3.

Fig. 19. Urino-genital system. Figure on the left (I) represents lower part and genital ducts. (K) lower part of ovarian sac; (J) spiral oviduct; (O) genital aperture; (I*) oviducal gland. Figure a little higher on the right (II), shows the details of aperture of oviduct (O) and urinary organ (R). × 5.

Fig. 20. Oviducal gland. Details of the glands that make it up. Approx. × 12.

*In the illustration given as T—General Editor.

PLATE VI

Figs. 1 to 14. *Vitreledonella richardi* Joubin

Fig. 1. Section of the liver. (G) Corpuscles with nuclei of different shapes; (H) hepatic cells; (O) large cells, stained; (R) connective tissue cells; (S) hepatic channels; (V) blood vessel. × 400.

Fig. 2. Transverse section of a branchial arc. (A) efferent capillaries; (B) brown cells; (D) distributing veins; (F) fibro-cartilaginous tissue; (N) nerve cells; (V) afferent vein of support (fibro-cartilaginous support). × 30.

Fig. 3. Section of a part of branchial gland. (C) gland cells; (D) lymphoid cells; (L) lacunae. × 350.

Fig. 4. Section through the middle of the entire branchial gland. (C) gland cells; (D) lymphoid cells; (L) lacunae. × 100.

Fig. 5. Funnel organ. Funnel split open to show the gland in the form of a transverse pad. (G) gland; (S) funnel. × 2.

Fig. 6. Transverse section showing the complete structure of the funnel gland. (E) glandular epithelium; (M) wall of the funnel. × 118.

Fig. 7. Transverse section of the funnel organ highly magnified. (Gs) surface gelatinous limiting layer; (Gp) basal or deeper gelatinous connective tissue layer; (H) small cells in these two layers; (M) long cells of the organ proper; (N) their elongate nuclei in the lower part; (O) mucous content of the cells; (S) highly stained spindle-shaped content of the cells. × 1126.

Fig. 8. Ventral view of the brain. (GC) cerebral ganglion; (GV) visceral ganglion; (O) esophagus; (Op*) optic nerve; (Olf) olfactory nerve; (Oph) ophthalmic nerve; (Pa) pallial nerve; (Vis) visceral nerve; (S) salivary duct; (V) vein. × 17.

Fig. 9. Side view of the brain. Letters as in Fig. 8. B_1, B_2, B_3 and B_4—brachial nerves of four arms; (GB) buccal ganglion; (GC) cerebral ganglion; (GV) visceral ganglion.

Fig. 10. Vertical section of the oviducal gland, slightly diagrammatic. (GC) glandular constituents; (Es) secretory epithelium; (M) secretory duct, glandular reservoir; (S) connective tissue space. × 30.

Fig. 11. Oblique tangential section of the oviducal gland. Letters same as in Fig. 10. × 30.

Fig. 12. Section showing the point where the ductules of elementary glandules reunite to form the secretory canal of the oviduct from the oviducal gland. Approx. × 40.

Fig. 13. Details of epithelial cells of secretory duct of one of the oviducal glands. × 130.

Fig. 14. Side view of a brachial sucker. (NB) brachial nerve; (P) small ganglion at the base of the suckers. Approx. × 25.

Figs. 15 to 17. *Graneledone verrucosa* (Verrill)

Fig. 15. Bifurcate tip of the arm. Approximately natural size.

Fig. 16. Hectocotylized arm. Approximately natural size.

Fig. 17. Hectocotylus. Approx. × 4.

**Note*. Marked On in the figure.

'II

T.

O.'

PLATE VII

Figs. 1 and 2. *Bolitaena diaphana* (Hoyle)

Fig. 1. Ventral side of statocysts. × 15.
(P) pedal ganglion; (O) statocyst, exposed; (O′) statocyst, intact; (C) auditory ridge; (S) round spot; (T) spot, leaf-shaped; (V) visceral ganglion.

Fig. 2. Dorsal side of the brain. (A, B, C and D) four swellings from which emerge the composite bundles of the optic nerve; (AO) aorta; (Gp) peduncular ganglion; (M, N, R, and S) swellings of the central group; (O) optic nerve; (Pa) pallial nerve; (U,V) swellings of the anterior group; (OE) esophagus; (O, I) olfactory nerve; (T) vein; (X) optic ganglion; (Z) visceral ganglion. × 19.

Figs. 3 to 6*. *Cycloteuthis sirventi* Joubin

Fig. 3. Dorsal side. × 1.4.

Fig. 4. Ventral side. Mantle split open to show the position of luminous organs. × 1.4.

Fig. 5. Luminous organs. × 4.5.

Fig. 6. Tentacular club. × 8.

Fig. 7*. *Megalocranchia hyperborea.*

Fig. 7. Details of funnel apparatus. × 12.

*In original these two numbers were respectively 4 and 6—Scientific Editor.

PLATE VIII

Figs. 1 to 11. *Cycloteuthis sirventi* Joubin

Fig. 1. Mouth and buccal-branchial membranes. × 8.

Fig. 2. Profile of mantle cartilage ridge. × 11.

Fig. 3. Mantle cartilage button corresponding to the ridge in Fig. 2. × 8.

Fig. 4. Front view of the funnel organ. × 5.

Fig. 5. Front view of the brachial sucker. × 60.

Fig. 6. Front view of the tentacular sucker. × 50.

Fig. 7. Details of the horny ring of the tentacular sucker. × 300.

Fig. 8. Ventral side of the gladius. × 3.3.

Fig. 9. Details of the posterior tip.

Fig. 10. Some luminous organs on the eyelid margin. × 20.

Fig. 11. Radula. × 60.

PLATE VIII

PLATE IX

Figs. 1 to 3. *Graneledone verrucosa* (Verrill)

Fig. 1. Ventral side. Reproduced from a photograph, slightly reduced in size.

Fig. 2. Side view. Reproduced from a photograph, slightly reduced in size.

Fig. 3. Dorsal side. Reproduced from a photograph, slightly reduced in size.

PLATE X

Figs. 1 to 3. *Heteroteuthis dispar* (Rüppell) Gray

Fig. 1. Ventral side. Reproduced from the watercolor sketch by Miss de la Roche. Approx. × 6.

Fig. 2. Side view. Reproduced from the watercolor sketch by Miss de la Roche. Approx. × 6.

Fig. 3. Dorsal side. Reproduced from the watercolor sketch by Miss de la Roche. Approx. × 6.

Figs. 4 and 5. *Pyroteuthis margaritifera* (Rüppell) var. *aurantiaca* (Joubin)

Fig. 4. Ventral side. Reproduced from the watercolor sketch by Mr. Chivot. Slightly enlarged.

Fig. 5. Dorsal side. Reproduced from the watercolor sketch by Mr. Chivot. Slightly enlarged.

PLATE X

PLATE XI

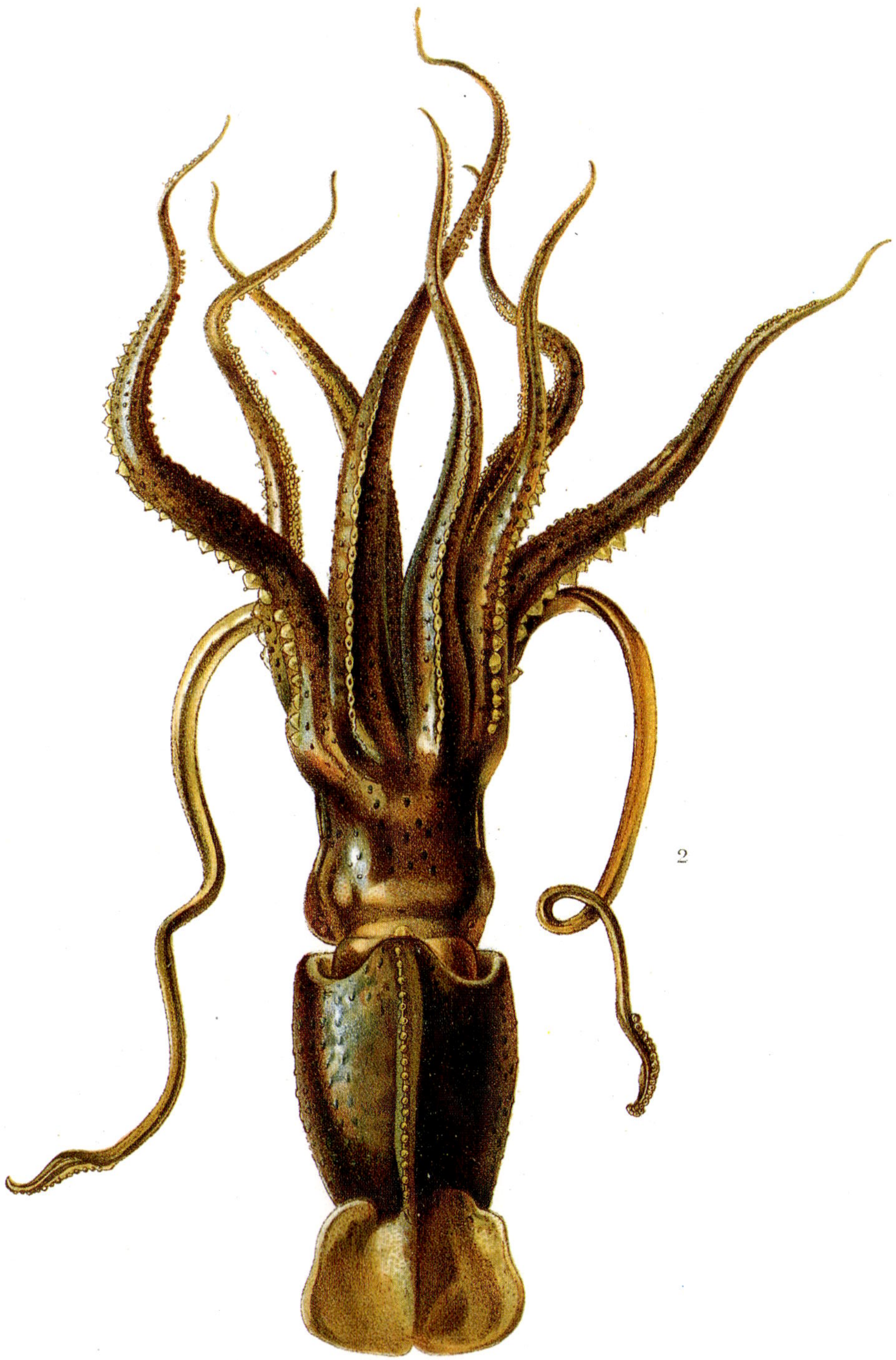

2

1

PLATE XI

Figs. 1 and 2. *Meleagroteuthis hoylei* Pfeffer

Watercolor sketch by Miss Vesque of a specimen preserved in sea water and formalin. Figures slightly reduced.

Fig. 1. Dorsal side.

Fig. 2. Ventral side.

PLATE XII

Figs. 1 to 5. *Meleagroteuthis hoylei* Pfeffer

Fig. 1. Cartilaginous teeth of the edge of the third arm. × 13.

Fig. 2. Part of the median crest on the skin of the dorsal side of the mantle. × 13.

Fig. 3. Asymmetry of the eyes. The eye on the right is smaller than on the left, surrounded by a complete circle of luminous organs disposed on the margin of the eyelid. × 3.

Fig. 4. Asymmetry of the eyes. The eye on the left is larger than on the right, surrounded by atrophied luminous organs, irregularly disposed on the margin of the eyelid. × 3.

Fig. 5. Part of the ventral skin of the mantle showing the complex luminous organ with large upper lens, and smaller frontal lens. (M) mirror; (LS) upper lens; (LF) frontal lens; (G) pigment sheath; (C) chromatophore of frontal lens. × 25.

Figs. 6 to 8. *Megalocranchia abyssicola* (Goodrich)

Fig. 6. Funnel organ as seen from the ventral side. × 8.

Fig. 7. Tip of the tentacle showing the terminal arrangement of the suckers. × 25.

Fig. 8. Median section of the visceral complex passing through the sagittal plane of the four luminous organs. (A) anus; (CB) branchial heart in the background; (IP) rectum; (R) ink reservoir; (U) urinary sac. × 10.

Fig. 9. *Bolitaena diaphana* (Hoyle)

Visceral complex seen from the back after lifting the covering membrane. (OE) esophagus; (Gs) ventral salivary gland; (J) crop; (E) stomach; (F) liver; (R) renal organ; (CB) branchial heart; (N) gonad; (V) ventricle. × 14.

PLATE XII

11

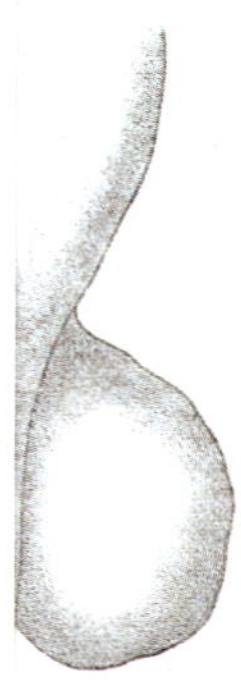

6

PLATE XIII

Fig. 1. *Bolitaena diaphana* (Hoyle)

Specimen dissected to show the general disposition of the viscera. (B) buccal bulb in the anterior, surrounded by chromatophores; (C) brain; (OE) esophagus; (Pa) pallial nerve; (GP) peduncular ganglion; (S) sinus; (MV) gilded or golden visceral mass; (X) optic ganglion. × 8.

Figs. 2 and 3. *Chiroteuthis lacertosa* Verrill

Fig. 2. Dorsal side of the animal, natural size, reproduced from a water-color sketch by Miss Vesque.

Fig. 3. Tentacular club, showing the arrangement of suckers and membranes. Approx. × 4.

Figs. 4 to 11. *Chiroteuthis portieri* Joubin

Fig. 4. Ventral side of the animal, natural size, reproduced from the water-color sketch by Miss Vesque.

Fig. 5. Dorsal side of the body of the same animal.

Fig. 6. Ventral view of the posterior end.

Fig. 7. Lateral view of the funnel showing its curvature, aperture, and funnel locking apparatus. × 2.

Fig. 8. Adhesive ridges situated on the inner margin of the mantle. × 13.

Fig. 9. Left adhesive mantle groove on the funnel pillar. × 13.

Fig. 10. Right adhesive mantle groove on the funnel pillar. × 13.

Fig. 11. A brachial sucker seen from its opening. Approx. × 75.

PLATE XIV

Figs. 1 to 4. *Megalocranchia abyssicola* Goodrich

Fig. 1. Ventral view of the entire animal. × 2.3.

Fig. 2. Mantle cavity dissected to show the visceral complex. In the center are four luminous organs between the gills; above these are the rectum and anus followed by the large sinuous vein; behind the gills are the two openings of the mantle respiratory cavity. × 5.

Fig. 3. Details of the luminous organs. (A) anus; (B) gills; (I) lower luminous organs; (S) upper luminous organs; (R) rectum. × 7.

Fig. 4. Side view of the tentacular club. × 12.

Figs. 5 and 6. *Brachioteuthis beani*

Fig. 5. Ventral view of the specimen; photographed after preservation in formalin. × 2.3.

Fig. 6. Tentacular club. × 6.

The present publication is deposited in the Oceanographic Museum of Monaco.